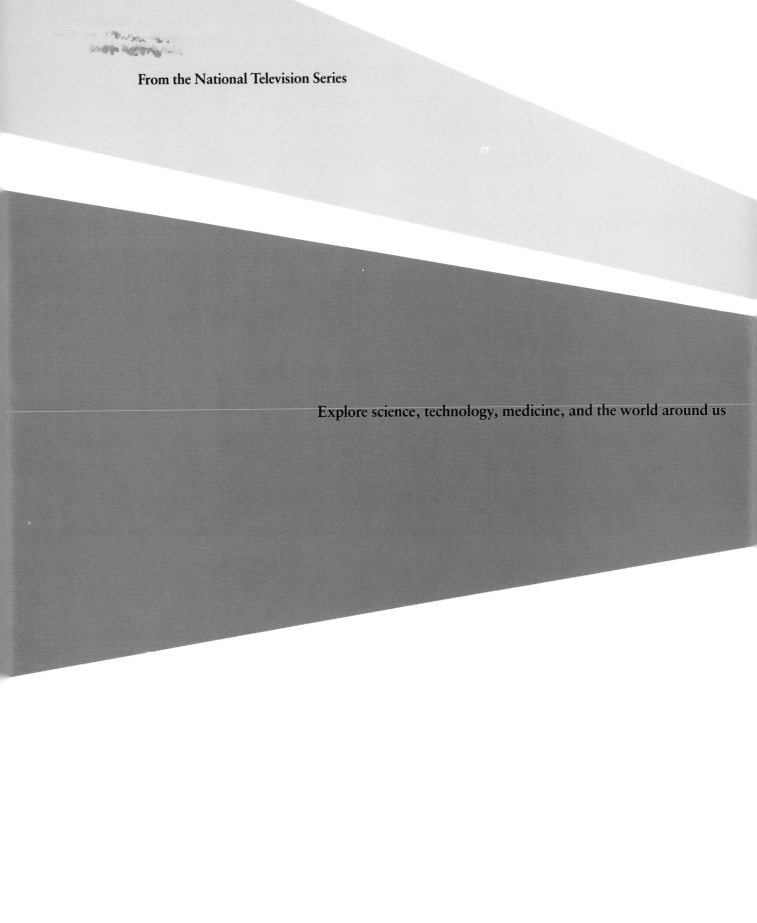

From the National Television Series

Explore science, technology, medicine, and the world around us

Television Producer
Twin Cities Public Television / KTCA Minneapolis–St. Paul, Minnesota

NEWTON'S APPLE

Publisher

General Communications Company of America Los Angeles, California

Contents

Behind the Scenes 1
- 2
- 3 Are We Alone in Space?
- 8 *Newton's Lemon* — Trick Hat
- 9 *Dead Ernest* — Why Do People Faint?
- 10 Fiber Optics: New Uses for Light
- 14 Dolphins

Behind the Scenes 2
- 16
- 17 The Unpredictable Curve Ball
- 19 *Newton's Lemon* — Car Park
- 20 Seeing Voices
- 21 *Dead Earnest* — Why Do We Get Goose Bumps?
- 23 Red-Tailed Hawks

Behind the Scenes 3
- 26
- 27 Learning to Deliver: The Space Shuttle
- 33 Clouded Leopards
- 34 *Newton's Lemon* — Car of the Future
- 36 Chasing the Common Cold

Behind the Scenes 4
- 38
- 39 Hypnosis: Your Attention Please!
- 42 Snakes
- 44 *Newton's Lemon* — Matchless Cigarettes
- 45 Why Is the Sky Blue?
- 47 *Dead Earnest* — Why Do We See Stars When We Are Hit on the Head?

Behind the Scenes 5
- 48
- 49 Video Game Fever
- 51 *Dead Earnest* — What Is the Funny Bone?
- 53 Holography: Taking Pictures in 3-D
- 58 Is There Any Way to Cut Onions Without Crying?
- 59 *Newton's Lemon* — Golf by Proxy

Behind the Scenes 6
- 60
- 61 Depth in Sight
- 65 Wolves
- 67 *Dead Earnest* — Why Do Our Ears Pop?
- 68 *Newton's Lemon* — Pocket Ski Lift
- 69 The Heart Hook-Up

As you read this book, bear in mind that scientists frequently disagree about scientific theories and explanations. In every case we have tried to present credible, well-considered explanations, but *not all* of the theories and explanations on all of the subjects.

Copyright © 1983 by General Communications Company of America. All rights reserved. No part of this book may be reproduced in any form or by any means without the prior written permission of the publisher, excepting brief quotations in connection with reviews specifically prepared for newspapers, magazines or television.

Newton's Apple — trademark pending. Property of Twin Cities Public Television, Inc.

ISBN 0-914761-00-5
Manufactured in the United States of America

Front cover photograph and all interior photographs not otherwise attributed were taken by Brian Paulson.

Library of Congress Cataloging in Publication Data
Main entry under title:

Newton's apple.

1. Science — Miscellanea. 2. Newton's apple (Television program) I. General Communications Company of America.
Q173.N45   1983   500   83-20514
ISBN 0-914761-00-5

Behind the Scenes 7
72
- 73 A World of Color Not Seen
- 77 The Tongue and Taste
- 79 *Dead Earnest*—Why Does Your Hand Fall Asleep?
- 82 Black Holes in Space

Behind the Scenes 8
84
- 85 What Is Digital Sound?
- 89 *Newton's Lemon*—Machine for Walking on Water
- 90 Bactrian Camels
- 92 Cooking the Good Egg
- 93 *Dead Earnest*—Why Do Teeth Chatter?

Behind the Scenes 9
94
- 95 The Robots Are Coming! The Robots Are Coming!
- 101 What Fat?
- 103 *Dead Earnest*—Why Do We Get Headaches from Eating Ice Cream?
- 104 Parrots

Behind the Scenes 10
106
- 107 Dressing for Space
- 112 Porcupines
- 114 Super Sight with Ultrasound
- 117 *Dead Earnest*—Why Do Bones Crack?

Funding for the Newton's Apple television series has been provided by **DUPONT Better Things for Better Living**

Proceeds from the sale of this book will help support public television.

Additional copies of *Newton's Apple* may be obtained by using the order form at the back of this book.

🍎 Foreword

What's a "Newton's Apple," anyhow? You've probably heard the tale of Isaac Newton's close encounter with an apple—how supposedly Newton conceived the idea of gravity when an apple fell on his head. Well, it's a great story, but it probably never happened. Sir Isaac's genius was expressed in the thought that gravity works not only on Earth, with apples, but everywhere in the universe, on the other planets and on stars. This was a revolutionary idea at the time, and it proved to be true.

Even though the story of Newton and the apple may not be true, it has been kept alive because it shows that by questioning even the simplest things around us, like the falling of an apple, we can uncover fascinating and enlightening ideas about the world in which we live. That's why we chose Newton's apple as the symbol and the title of our program.

We wanted Newton's Apple to be a different kind of science program where viewers and the audience get involved in the questioning. We wanted to show the excitement of science and that to understand science you don't have to be an Isaac Newton or an Albert Einstein. So we made Newton's Apple fun. We invited viewers to mail in their questions and, to answer them, we brought in living experts with lively demonstrations. That's what I like most about this program: It brings science out of the laboratory and into your living room.

If a genius like Newton wasn't afraid to ask questions, why should we hesitate? Of course we won't discover gravity, but there's a good chance we might uncover something just as meaningful: the desire within all of us to know more about our universe.

Ira Flatow

Behind the Scenes 1

NEWTON'S APPLE

■ The day we worked with the dolphins they so enchanted us that we forgot they were animals and came to think of them as friends. When Ira got into the water with them he said he could feel the dolphins "zapping" him. Nancy Gibson, our expert from the Minnesota Zoo, explained that they were "echolocating" Ira. She said they send sound waves through the water to check you out on their personal sonar systems. ■ It was amazing how responsive the dolphins were to their trainer. When he was around, they behaved like two obedient school children. ■ The segment was one of the most difficult for us to film because we wanted to make it informative and entertaining while ensuring that the safety of Ira, Nancy and the dolphins would not be threatened. ■ One tense moment came when the underwater photographer got into the water dressed in what looked like an old-fashioned diving suit, outfitted with a big bubble helmet and protruding oxygen hoses and sound wires that enabled him to communicate with our production crew. The dolphins didn't know what to make of him. They zoomed right at him and veered away only at the last second. They were so curious and distracted by him that for a while they did not perform their usual behavioral acts. We felt sympathetic because he did look weird, even to us!

Billions of years from now our sun, then a distended red giant star, will have reduced Earth to a charred cinder. But the Voyager record will still be largely intact, in some other remote region of the Milky Way galaxy, preserving a murmur of an ancient civilization that once flourished—perhaps before moving on to greater deeds and other worlds—on the distant planet Earth. —Carl Sagan

Are We ALONE In SPACE?

The Andromeda Galaxy, one of our closest neighbors.

This toy radio transmitter is just like the one that E.T. used to phone home. The homemade galactic CB radio was invented by Henry Feinberg for the movie *E.T., The Extra-Terrestrial*. Henry was once a production assistant to Don Herbert, a man better known as Mr. Wizard. Henry's final design was a beacon transmitter that could operate unattended in a forest clearing, yet be capable of directing a pinpoint microwave signal into space. The transmitter really doesn't work, being a version of Walt Disney's "plausible impossible" concept. This concept can best be explained by analogy to the cartoon character who runs off the side of a mountain and does not begin to fall until he realizes that there is no longer any ground under him.

E.T. the loveable little extraterrestrial created by Steven Spielberg, took the country by storm because he touched a universal longing in us that is as old as humanity itself: the desire to know our universe and to understand our place in it.

Are we the only intelligent beings in a gigantic universe? Or can it be that other civilizations exist—millions, billions, or even trillions of miles away? Human beings have been speculating on these questions since the dawn of civilization. We have done a good job of observing and describing our celestial backyard—the solar system and the Milky Way galaxy—yet we still don't know whether intelligent life exists beyond our own planet.

Only in the past few years have scientists begun to take an active interest in extraterrestrial communication. With radio technology capable of transmitting and receiving messages in outer space, prospects of communicating with intelligent life no longer seem like science fiction. E.T. used a toy radio transmitter to phone home, and we are using complicated audio and visual apparatus, but the message is similar. We want to let others know who and where we are.

Pioneers 10 and 11, NASA's first interstellar space vehicles, carry plaques inscribed with the visual messages shown in

We want to let others know who and where we are

🍎

the illustration. A man and a woman stand in front of a drawing of the spacecraft, so that the viewer can look at the craft itself and know what size we are. The two circles represent the hydrogen atom in its two energy states. The connecting line between the two circles, along with the digit 1, indicates that the time interval between the transition from one state to the other is to be used as the fundamental time scale. The pulsar map to the right of the man and woman pinpoints the location of our solar system in relation to 14 pulsars, whose precise periods of pulsation are recorded in binary numbers (1 and 0). Underneath these illustrations is a schematic map of the solar system indicating that the spacecraft were sent from the third planet out from the sun. The spacecraft are shown to pass Jupiter and then travel out of the solar system.

Voyagers 1 and 2, launched in 1977, carry similar drawings, plus a kind of phonograph record that has been encoded to present sights and sounds of the Earth, as well as greetings in 60 languages.

We asked Dr. Timothy Ferris, a professor of astronomy at the University of Southern California and producer of the two-hour phonograph record called "Sounds of Earth," to visit with us on Newton's Apple and tell us more about intergalactic communication.

Ira: Timothy, what are the chances of our finding intelligent life out there with this message-in-a-bottle technique?
Timothy: Well, if there were, say, 10,000 advanced civilizations in the galaxy, we

Ira with
Dr. Timothy Ferris

think the odds would be comparable with the odds that your aunt in New Zealand would find a bottle that you threw into the Atlantic Ocean. It might come to pass, but chances are very slim. The diagram plaques affixed to the Pioneer spacecraft when they were launched in 1971 and 1972 did not convey much information. They were limited to just one picture. By the time NASA built the two Voyagers, space scientists had realized that they could put a lot more information on a phonograph record, so we did. The record, a 12-inch copper disk, contains samples of music from various cultures all over the world, natural sounds of surf, wind, thunder, birds and whales, and over 100 photographs.

Ira: How far from the Earth will these spacecraft be able to carry the message?

Timothy: Well, consider the satellite called LAGEOS, an acronym for Laser Geodynamic Satellite. This small satellite was launched in 1974 mainly for the purpose of measuring the drift of Earth's continents by recording the separations in the distance between Earth-based laser beams being bounced off LAGEOS. The estimated lifetime of LAGEOS is eight million years, and NASA decided that this should be the first spacecraft to carry a greeting card to our descendants. The two Pioneers and the two Voyagers will travel through our galaxy for over 100 times that long—a billion years or more. A much more efficient way to communicate between stars is by radio waves. These waves travel at the speed of light, which means that radio waves travel much farther in five seconds than our spacecraft travels in one day. With this method astronomers have broadcast our messages to the stars from the world's largest radio telescope at Arecibo, Puerto Rico. So, even though the Voyagers will travel a vast distance—they will orbit the massive center of the Milky Way galaxy once every quarter billion years—the messages sent by radio telescopes will travel much farther and faster.

LAGEOS carries a greeting to our descendants on Earth eight million years from now

Ira: How do we send a diagram by radio waves?

Timothy: Just the way we send television pictures. We encode the message in two numbers—ones and zeros—a system called binary numbers. If you convert these numbers into pictures you end up with images like those of the plaques on the spacecraft. That message was created by Professor Frank Drake, one of the founders of SETI (Search for Extraterrestrial Intelligence) and the first astronomer to listen for extraterrestrial radio messages. This particular message shows the planets of the solar system, atoms of carbon and oxygen, and the structure of the DNA molecule, the basis of life on earth.

Ira: How about this message on the record cover of the "Sounds of Earth" that you produced to go along with the Voyager missions. What can you tell us about it?

Timothy: Well, since the real message is on the phonograph record, the message on the cover is mainly about how to operate the record. The drawing in the lower lefthand corner is the pulsar map previously sent as part of the plaques on Pioneers 10 and 11. And electroplated on the record's cover is an ultra-pure source of uranium-238, from

Continued on page 9

FAR OUT

The first television broadcast has traveled far enough to have reached about four hundred stars.

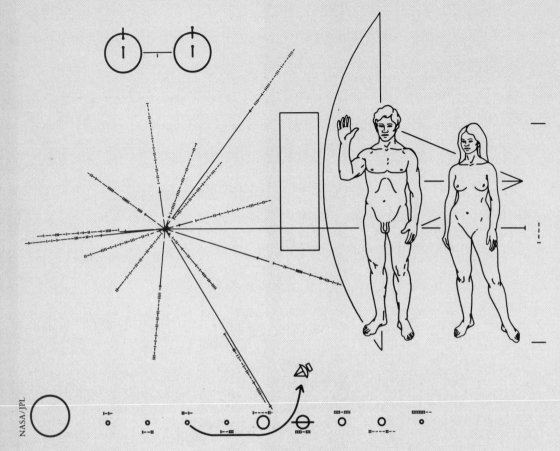

The Pioneer spacecraft, which are the first objects made by humans to escape from the solar system into interstellar space, carry this pictorial plaque. It is designed to show scientifically educated inhabitants of some other solar system —who might intercept it millions of years from now—when the spacecraft were launched, from where, and by what kind of beings.

When the recording was being produced, the President of the United States was Jimmy Carter, who sent the following message: This is a present from a small distant world, a token of our sounds, our science, our images, our music, our thoughts and our feelings. We are attempting to survive our time so we may live into yours. We hope someday, having solved the problems we face, to join a community of galactic civilizations. This record represents our hope and our determination, and our good will in a vast and awesome universe.

Voyager

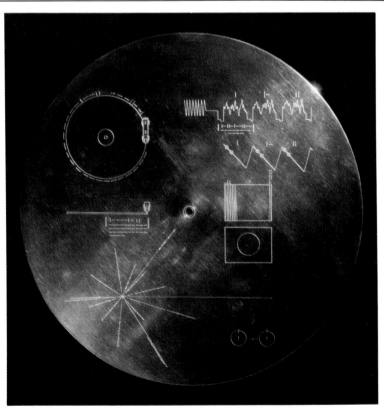

6 ■ NEWTON'S APPLE

Reaching Out...

Were we to locate but a single extraterrestrial signal, we would know immediately one great truth: that it is possible for a civilization to maintain an advanced technological state and not destroy itself. We might even learn that life and intelligence pervade the Universe . . . [as well as] all sorts of scientific results, ranging from a valid picture of the past and the future of the Universe through theories of the fundamental particles to whole new biologies. Some conjecture that we might hear from near-mortals the views of distant and venerable thinkers on the deepest values of conscious beings and their societies! Perhaps we will forever become linked with a chain of rich cultures, a vast galactic network. Who can say? — from a SETI report

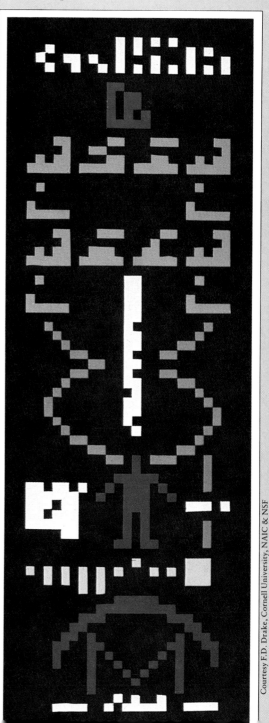

Above, the radio telescope at Arecibo, Puerto Rico. On the right is astronomer Frank Drake's binary message to the cosmos. It consists of 1679 ones and zeros (or *bits*). The number 1679 is the product of two prime numbers, 73 and 23; by arranging the ones and zeros in 73 rows of 23 bits each and assigning black squares to zeros and white squares to ones, extraterrestrials can obtain the picture shown here. Sophisticated decoding then yields the numbers 1 through 10 on the top row, followed by the atomic numbers of the five essential elements of terrestrial life: hydrogen, carbon, nitrogen, oxygen and phosphorus. Next come 12 blocks of numbers that give the chemical formulas for DNA, which is symbolized by the helix seen through the middle of the picture. Other features include a stick figure of a human being with height and weight expressed in wavelengths, the Earth's population, and our solar system with Earth slightly displaced. The bottom diagram shows the Arecibo telescope that transmitted the message. Sent seven years ago, Drake's message has already passed several of the nearest stars.

This giant tracking antenna, 210 feet in diameter, is located in Goldstone, California. It is part of NASA's Deep Space Network of stations where the most sophisticated radio receivers in the world maintain 24-hour radio contact with planetary and interplanetary space vehicles. To comb the cosmic haystack for extraterrestrial intelligence, Paul Horowitz and other space scientists will hook up this antenna to a unique radio receiver, called a multichannel signal analyzer, that can divide a radio signal into 8,000,000 channels and examine them all simultaneously for signals.

TRICK HAT

Now here's a hot idea on how to cool a fevered forehead. Horatio Castervilt invented a skimmer that he thinks will drop the temp a few degrees. A battery-operated fan does the trick. And it works with a powerful little motor on the top of the hat. But what if his fan gets caught in his hair? Well, toupee or not toupee....

8 ■ NEWTON'S APPLE

Continued from page 5

which an extraterrestrial recipient could calculate the time elapsed since a spot of uranium was placed aboard the spacecraft.
Ira: What do you think is the probability of someone sending a radio or some other kind of message back to Earth?
Timothy: It's impossible to say. If other people, other creatures, other intelligences do receive our messages, then they might send back complex messages lasting for months and years—whole encyclopedias of information—by radio.
Ira: But how do we know what to listen for?
Timothy: At present our radio telescopes can detect natural radio energy from millions of light years away. This enormous range embraces billions of stars. A radio telescope is a large dish that collects radio energy and then amplifies it. A big radio telescope like the ones we use to listen to the natural radio noise of the cosmos is extremely sensitive. To appreciate how sensitive, we can compare it with the flame from a match. There is more energy emitted by the flame in a few seconds than has been collected by all the world's radio telescopes in all the decades that radio astronomy has been listening to the cosmos. So a very faint signal could be received from very far away, if you knew where to look and what frequency to listen to.

Paul Horowitz, a physicist at Harvard University, has constructed an advanced radioastronomical receiving system that permits us to listen to over 130,000 adjacent radio channels all at once. This should

---

Someone on Alpha Centauri could receive the Newton's Apple broadcast about four years from now

---

greatly increase the efficiency of SETI programs. This new system can perform more SETI searches in one minute than Frank Drake's 1960s system could have done in a hundred thousand years. And NASA is working with an array of telescopes that will eventually receive more than eight million separate channels. In addition, Japan and the Soviet Union are constructing their own apparatus for SETI research.
Ira: How far away is the closest potential receiver and sender of messages?
Timothy: Our nearest star is Alpha Centauri, a little more than four light years away. If anyone were there to receive this broadcast of Newton's Apple, they would see the show in four years and a few months from now.

SETI's plans for future research are on a grand scale. Ever since 1975, when the powerful Arecibo radio dish tuned in five close galaxies and Earth's 200 nearest stars, space scientists have imagined telescopic arrays of even greater proportions and power.

Cyclops, a mammoth array of 1500 radio dishes, each 100 meters across, is one of their dreams. Beyond that, there is a proposal for a huge orbiting receiving dish that would be launched into space by shuttle to beam information back to Earth. With such visionary plans, the probability of a close encounter becomes closer and closer—though it may be farther and farther away. 🍎

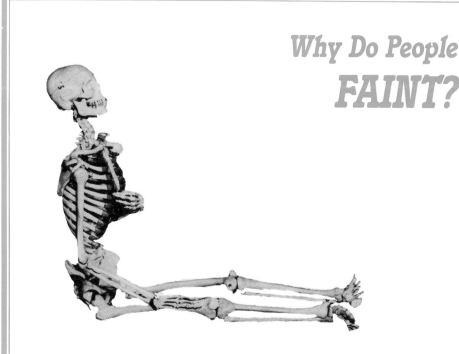

Why Do People FAINT?

Ira: Here's Dr. Janet Serie with her friend Dead Earnest, and it looks as if he's brought along his heart and a couple of blood vessels. I always thought Dead Earnest didn't have a heart. What is all this about, Jan?
Jan: Dead Earnest wants to help me to explain why people faint. Those are the carotid arteries that carry blood to the brain. You see, fainting occurs when the brain is deprived of oxygen from the blood and you become unconscious. Fainting is common with people who have low blood pressure, because the blood is too slow in getting to the brain.
Ira: Is fainting dangerous?
Jan: Not at all. Even if you pass out completely. That is nature's way of getting your brain down to the level of your heart so it doesn't have to work so hard. The only danger in fainting is that you may injure yourself in falling. So if you get a sort of woozy feeling and think you are going to faint, just sit down and put your head between your legs so that the blood can reach your brain easily. You should revive quickly.
Ira: What about soldiers who faint while standing at attention?
Jan: Their blood pools in their legs, so less of it gets to their brains and they just pass out. Victorian women were known to swoon dead away when someone came courting. That is because they wore their tightest corsets on such romantic occasions, and it prevented the blood in their lower bodies from returning to their hearts.
Ira: You mean it didn't have anything to do with the guys at all? Oh boy, did you just destroy a myth!

FIBER OPTICS

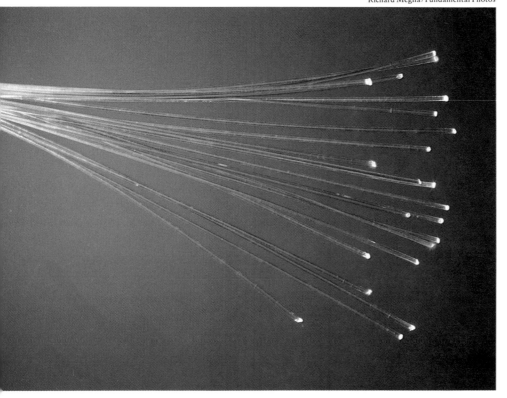

Richard Megna/Fundamental Photos

Light is being projected through these long, thin fibers by a laser beam. What makes these fibers unique is their ability to funnel the light along the inside of the fibers so that very little escapes through the sides. That is why the fibers appear bright at the tips and not along the sides.

Tiny glass fibers as fine as one strand of your hair are revolutionizing our information age. The science of using light pulses instead of electrical signals to transmit information is known as *fiber optics*. Ten years ago it was only a dream, but today light pulsed along a glass fiber by a miniature laser can transmit nearly limitless amounts of information in the form of voice signals, video signals, data and graphics.

Optical fibers work on a simple principle. Just as a copper wire conducts electrons, optical fibers conduct photons, or discrete packages of light. A laser small enough to fit under your fingernail projects light through hair-thin glass fibers. These fibers can guide the light waves over long distances at extremely high speeds. Because one of the basic properties of light is that it generates

Optical fibers can transmit all of Webster's Unabridged Dictionary in six seconds

higher frequencies than electricity, a lightwave system can transmit far more information than an electrical one.

Telecommunications of all kinds are faster and more efficient where the conventional method of transmitting electrons through metal (usually copper wire) has been replaced with transmission of photons through glass. Optical fiber cables are a tiny fraction of the diameter (four one-thousandths of an inch) and the weight (only 1 percent) of copper cables, so they are easy to handle and take up far less space. An optical cable with a diameter roughly the size of your finger can transmit as much

> The technological advances that made [progress] possible are clearly identifiable: the transistor, which led to the computer technology that enables us to store and process vast amounts of information; the laser, which made it possible to use the light wavelengths of the electromagnetic energy spectrum for communication purposes; and the optical fiber, a transmission medium compatible with the laser and capable of carrying vast amounts of information at very high speed. —Dr. C.K. Kao, ITT

NEW USES FOR LIGHT

data as a standard 2¼-inch-diameter coaxial copper cable with 20,000 wires.

Optical fibers can transmit information at fantastic speeds. A recently installed lightwave system between New York City and Washington, D.C. can transmit the entire contents of Webster's Unabridged Dictionary—more than 2700 pages—in only six seconds. At present, one pair of glass fibers as thin as a hair can carry 1300 simultaneous telephone conversations, whereas no more than 24 can be handled by conventional copper cables.

Because light encounters little resistance from the ultrapure fibers through which it travels, light pulses can travel much farther than electrical signals without being reamplified by special boosters. In practical terms, this means that lightwave systems require a signal regenerator only every 10 miles instead of one every 300 to 6000 feet, which is required by copper cables.

Optical fibers are resistant to corrosion, electromagnetic interference, lightning,

No more catastrophic sparks in petroleum refineries

🍎

wiretapping and other problems associated with copper wiring. Since lightwave cables don't conduct electricity, they can carry signals between two points that have a large difference in voltage. Another advantage to the nonelectric nature of optical fibers is that when an environment is potentially explosive, as in petroleum refineries, optical fibers don't create catastrophic sparks.

In addition to the fiber itself, the other light components required for fiber optics communication are a light emitter (the la-

Step-index fiber — Light bounces in zig-zag path

Graded-index fiber — Light curves in gentle path

Monomode fiber — Light moves in straight path

Equivalent capacity: copper and optical fiber

Courtesy Corning Glass Works, Corning, NY 14831

Ira with gastroenterologist Dr. Christopher Gostout of the Mayo Clinic

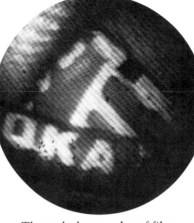

Through the wonder of fiber optics we get a look at the contents of Ira's pocket: keys, cough drops, coins (you can tell the difference between a penny, a nickle and a dime) and Bazooka Bubble Gum.

ser), digital encoders, a light detector (usually called a photodetector) and the links for the fibers and lightguides.

To grasp how the system works, consider what happens during an ordinary telephone call. Your words are first transformed into an electrical signal. This signal is scanned by a digital encoder, which reduces it to a series of ons and offs. The light emitter transmits the digital ons as a pulse of light and the offs as the absence of a light pulse. The light then travels through the lightwave cable to its destination, where it is received by the photodetector, amplified, and fed into another electrical signal. The person you are calling hears this signal and is able not only to receive the words you have spoken, but also to pick up the tiniest nuances that give clues to your state of mind. Just one detector connected to an amplifier can receive as many as 90 million bits of information each second. Astonishing as that is, rapid advances in the technology of lasers and lightguides will soon permit transmittal of *billions* of bits of information each second.

In the early years of fiber optics technology, medical applications predominated. Optical fibers are flexible, which means light inside the fibers can bend, and

Physicians can look around inside the human body

so they are ideal for probing the nooks and crannies within the human body. Flexible endoscopes, instruments for looking around inside the body, are invaluable aids to medical diagnosticians. With an endoscope, a physician can look directly into your gastrointestinal tract to see if you have an ulcer or some other problem. If the endoscope is also equipped with a tiny camera, it can be rotated in the stomach in order to take a series of pictures. Your physician can then compare the appearance of your stomach lining with others known to be healthy. If an ulcer is present, the endoscope can remove a piece of tissue for examination in the laboratory.

Though medicine and telecommunications led the way in developing applications for fiber optics, others quickly followed. Glass fibers now assist in seismic exploration of the earth and its oceans, oil well logging, coal mining, NASA's space shuttle, guidance systems for tactical missiles, optical gyroscopes, TV image intensification, industrial control systems, vehicular con-

CHICKEN CHAMP

A chicken does not lay an egg every day. The average hen lays about 240 eggs a year. The record was set by a chicken in New Zealand that laid 361 eggs in one year.

Optical fibers bend light. This property is useful for looking inside the body and lighting organs that doctors want to examine. The technical term for such a probe is *endoscope.* One endoscope is the gastroscope, used to see the upper gastrointestinal tract. Also doctors can use a forceps at the end of the gastroscope to take samples of tissue with little or no discomfort to the patient.

The laser has been used with optical fibers to perform surgery in obstructed places, such as arteries.

Courtesy Corning Glass Works, Corning, NY 14831

trol systems, power station controls and robots.

Over the next few years lightwave cables will link the major population areas of the world. Much of the Northeast corridor and California have already been hooked up, and a transcontinental system will probably be in service by the year 2000. Other ambitious fiber optics projects have begun in France, Japan, Canada, West Germany and Britain. The first transatlantic

Coast to coast card games

lightwave cable is expected to be in operation very soon.

Closer to home, tomorrow's telephone will become a key instrument in a varied communications and information center. We will be able to call up data storage banks, libraries, newspapers and magazines and automatically receive pictures and printed information about news events, fashion, sports, travel, books, or any other subject of interest to us. We may use our phones to vote in elections, to bank by credit cards, to make and charge purchases from department stores or to play poker coast to coast using video monitors. So get ready to ante up!

NEWTON'S APPLE ■ 13

Dolphins

John Perrone/Minnesota Zoological Garden

The ancient seafarers were as fascinated by the dolphins as we are today. So much did they admire them, in fact, that some ancient peoples believed these animals held the souls of sailors drowned at sea. Aristotle himself studied both dolphins and porpoises as they frolicked in the warm seas of the Aegean and Mediterranean. He decided correctly that they are warm-blooded mammals, not fish.

Albert C. Jensen, in his book *Wildlife of the Oceans*, relates what it is like to encounter dolphins at sea. "It is an exciting experience to see dolphins riding the bow waves of a vessel at sea—an experience already described by Greek sailors nearly 3000 years ago. On one occasion I saw a trio of these animals riding the bow waves of our research vessel, easily keeping pace with our speed of 12 knots. They move effortlessly, and it is said that they are able to swim in bursts of up to 25 knots, obviously having learned to take advantage of the hydrodynamics that operate as the bow of a ship cleaves the sea. Soon the sea virtually erupted with the graceful animals from horizon to horizon. Some took turns riding the bow waves; others merely swam along in the arcing movement called *porpoising*. Still others engaged in a marvelous ballet, leaping 3 meters or more into the air and sometimes twisting as they rose. Some returned smoothly to the waters, while others fell back on their sides, causing a great splash. It looked as if the dolphins were just having great fun, leaping and splashing like children playing in a pond; but as marine scientists, our judgment told us they were probably trying to dislodge parasites from the skin."

About one hundred species of whales, dolphins, and porpoises belong to a group of mammals called *cetaceans*. The single family of toothed whales is called *Delphinidae* and includes dolphins, porpoises, killer whales, pilot whales, and many other small species. Though they appear to be fish, these animals are mammals because they are warm-blooded, bear live young and breathe air. Their nostrils (blowholes) are positioned on the highest part of the head and are directly connected to the lungs. These mammals don't breathe through their mouths.

From studying the evolution of biological life, we know that many living things originated in the seas and later moved onto the land. Curiously enough, the great whales and dolphins returned to the sea after having lived on land for a few million years.

Playing—or scraping barnacles?

The common dolphin is cosmopolitan; it is found both in temperate and tropical seas and occasionally even in fresh water. Some dolphins make their way up river estuaries in search of fish to eat. Dolphins usually travel in groups of twenty or more individuals, but on occasion a hundred dolphins will swim together.

Some of the popular notions of dolphin intelligence originated because the animals are extremely, and almost constantly, vocal. Dolphins orient themselves by *echolocation,* a kind of built-in sonar system. They are capable of two kinds of sounds. A specialized mechanism in the nasal passages, just below the blowhole, enables them to emit short, pulsed sounds. These sounds, called *clicks*, can be produced in such rapid succession that they sound like a buzz—or even a ducklike quack. The clicks are actually produced by their echolocation apparatus, which enables them to detect obstacles, other dolphins, and even tiny bits of matter in the water.

Deeper in the respiratory system, presumably in the larynx, dolphins produce a high-pitched whistle or squeal, which is ca-

A rare sight. People almost never get this close to dolphins. Despite the many human-meets-dolphin stories, dolphins hardly ever approach swimmers in the wild. In Shark Bay, on the western coast of Australia, these dolphins swim into the shallows to frolic with waders. The children offer them small fish, which they accept but then drop to return for more. Both sides seem to be enjoying themselves, although there's no telling what the dog has on its mind. Despite their friendliness, dolphins can unintentionally but seriously harm swimmers, so in normal circumstances dolphins should not be approached.

From *The Dolphin's Gift*, Elizabeth Gawain, Whatever Pub.

pable of rapid pitch changes. The whistles differ from the clicks in being essentially single tones, while each click simultaneously comprises a broad spectrum of frequencies, including many in the ultrasonic range. Dolphins appear to use the whistles to communicate a certain emotional state, and this influences the behavior of the others. The squeals usually denote alarm or sexual excitement. Despite many efforts to demonstrate it, there is no convincing evidence that dolphins possess a communication system that could be considered a language. Even so, many researchers keep trying to learn how to communicate with them.

Ira: Nancy Gibson and I are here at the Minnesota Zoo to take a look at the dolphins. Nancy, these dolphins are considered to be very intelligent mammals, aren't they?
Nancy: Very smart. Their brain is actually bigger than ours.
Ira: What do they do with a brain that big? Why does nature give them a brain like that?

Whistles signal emotions and squeals reveal alarm

Nancy: Their brain is highly developed for acoustics. They're really a great hearing machine! They use echolocation, which is the ability to send sound at various frequencies. The sounds they send out echo back to them and give them information.
Ira: Like submarine sonar, or a bat?
Nancy: Yes. And with their sonar they can determine the size, shape, texture, and distance of objects or other animals. They are even able to increase the sound frequencies to the point where they can stun their prey by giving it a concussion. Let's take a swim with them. I should say, Ira, that dolphins are extremely popular with all visitors to the zoo because they are so graceful and seem so friendly, but the same precautions apply with all wildlife. It would be foolish to jump in the ocean with dolphins, as we're doing here, and expect them to play with us.
Ira: I was just touching her skin. How can we describe it to the people at home?
Nancy: To me it feels a little like a peeled, hard boiled egg.
Ira: I won't tell her you said that.

Behind the Scenes 2
NEWTON'S APPLE

- Would Ron Guidry, a busy pitcher for the New York Yankees, be able to come on the show and teach Ira how to throw a curve ball? The Yankees were in town to play the Minnesota Twins, and Ron was scheduled to pitch Friday and shoot the TV segment on Sunday. But if the Yankees switched Ron to Saturday, his arm would be too sore to pitch for us on Sunday. ■ We got lucky and Ron pitched on Friday, as scheduled. However, when we went over to the Metrodome on Saturday Ron seemed surprised to see us. We had made all the arrangements with Ron's agent in Louisiana. Had he forgotten to tell Ron? Panic struck. ■ After talking a bit, Ron did remember. A real pro, he quickly warmed up and, as we shot the segment, gave us lots of information on a subject he knows very well. ■ Ira still has a little trouble with his curve ball.

16 ■ NEWTON'S APPLE

The Unpredictable CURVEBALL

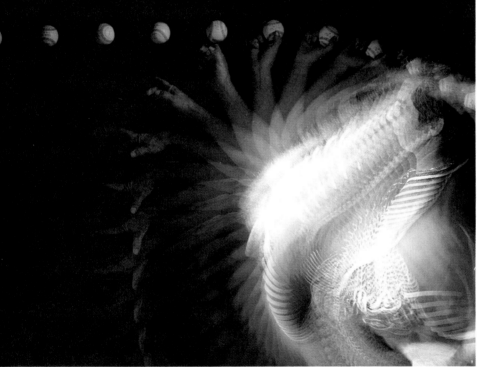

Gordon Gahan/Prism & Science 82; stroboscopic lighting by C.E. Miller/M.I.T.

> It doesn't look like the ball starts breaking until about 10 feet away, which can make a hitter flinch 'cause he's never too sure if the thing's a fast ball or a breaking ball. It's a matter of hanging in there. Some players can, some players can't. The ones that can't have other jobs by now. —Joe Nolan, catcher

Since the Civil War baseball enthusiasts, not to mention pitchers and batters, have been mystified by the behavior of the curve ball. This kind of pitch has gone by many names—drop ball, hook, outshoot—as people struggled to describe its action. After seemingly interminable debates about whether the curve ball actually curves at all, the aerodynamicists of baseball are turning their attention back to the century-old question of *where* the ball curves.

Our guest expert Ron Guidry, the breaking-ball hurler for the New York Yankees, succinctly summarized the phenomenon.

Ron: A good curve ball is something that comes over, falls in.
Ira: You're also well known for your slider pitch. What's the difference between a slider and a curve ball?
Ron: A curve ball probably has a bigger curve because its slower speed gives it more time to break. A slider has a short, downward break. It was developed to appear to be a fast ball so it would throw off the batter's timing.

Even though pitchers and batters seem to agree that the curve ball curves just a few feet from the plate, physicists disagree. They say that the ball curves constantly, so that if there were no gravity or obstructions, the pitcher could catch his own pitch.

Lyman J. Briggs, former director of the National Bureau of Standards, contended that the baseball's 216 stitches of red cotton pull a thin layer of air around them as they spin. Briggs determined that a curve ball has a top spin, which means that the air flows faster around the bottom than the top. Because the bottom air travels faster, the air pressure is lower. This means that the ball is pushed down by the combined effects of higher air pressure on top of the ball and lower pressure on the bottom. Briggs discovered that a major league curve ball rotates approximately 18 times in the half second that it travels to the plate and that it can curve about 17 inches.

Guidry: A good curve ball comes over, falls in

NEWTON'S APPLE ■ 17

Ira with Dr. Richard Brandt

With the aid of computer graphics we can see various ball trajectories and compare them with that of a curve ball. And with a wind tunnel made of a cylinder containing smoke, we can simulate the motion of wind around a baseball as the ball spins through the air.

We wanted to understand more about the science of the curve ball so we invited Dr. Richard Brandt, a professor of physics at New York University, to enlighten us.

Ira: What is there about the spinning baseball that makes a curve?

Dr. Brandt: Let's look at some computer graphics and see where various ball trajectories can go.

Ira: What's that straight black path?

Dr. Brandt: That's how a ball would travel if there were no gravity and no spin.

Ira: What about this green line that curves down a little?

Dr. Brandt: That shows the effect of gravity, making the line dip.

Ira: And what about the red trajectory?

Dr. Brandt: The red trajectory shows what happens when the spin and gravity are acting together. The spin causes further downward force, pulling the ball down and increasing the dip at the end of the trajectory.

Ira: There is quite a variation among the three. The spin really does impart that curve, doesn't it?

Dr. Brandt: Absolutely.

Ira: Now, how does it do it? That's the question we really want to ask.

Dr. Brandt: We'll use the wind tunnel to answer that one. When the ball isn't spinning, the smoke in the tunnel is uniformly distributed over the top and bottom of the ball. But when we spin the ball, look at the difference. The net effect is that the air on the bottom is moving faster than the air on the top. A principle in physics, called Bernoulli's Theorem, states that the greater the fluid's speed of motion, the less the pressure. The speed is greater on the bottom, so the pressure is less. When there is more pressure on the top than the bottom, there

CURVE AHEAD

A major league curve ball rotates about 18 times in the half-second it takes to get to the plate, and it can drop as much as 17½ inches.

is a downward movement caused by this pressure difference.

Ira: So the pressure on the top pushes the ball down, and then the ball follows a curved path. That makes sense. Now, if I were a smart pitcher, could I take advantage of that effect by doing something to the ball?

Those red stitches hold the secret

Dr. Brandt: Yes, you could. The threads on the ball are responsible for most of the friction between the air and the ball. So if you throw the ball in such a way that all four seams revolve during each revolution of the ball, then there will be more friction and the effect will be greater.

Ira: I know that a fast ball spins also. Why doesn't a fast ball take a dive?

Dr. Brandt: Because a fast ball spins in the opposite direction. The greater pressure on the bottom keeps it from falling as fast as it would if it fell simply from the force of gravity.

Ira: What happens if a ball doesn't rotate at all?

Dr. Brandt: That's called a knuckle ball. Rotation is so slight that, as the ball travels, the seams come up in different ways, which causes deflection upward or downward, left or right, in a kind of erratic way.

Ira: Now that sounds like my kind of pitch!

CAR PARK

With new cars getting bigger and parking spaces getting smaller, at Piedmont, California an inventor has developed something to soothe the motorist's headache by putting the spare tire to work. He calls this device the *park-car* and says it can be installed on any model. Taking power from the driveshaft, the spare tire swings the rear end into the clear. Then you just retract the spare, back into the street, and away you go! It's handy for parking in inaccessible garages, too. Even the worst driver can make it into the garage without denting a fender. With the aid of fifth-wheel driving, the car can turn a complete circle within its own radius. Finally, parking can be a breeze!

Seeing Voices

If someone tells you "You sound just like your mother" you can safely reply "I certainly do not." Through the use of a sound spectrograph it has been conclusively determined that each person's voice is unique. Like your fingerprints or your handwriting, your voice pattern is entirely individual, and now there is a way to take a picture of it.

The sound spectrograph responds to the complex sound waves generated by your voice by measuring each frequency in relation to the hundreds of others. It then displays them on paper as a series of vertical and horizontal bars commonly known as a *voiceprint*.

You don't sound just like your mother

Your vocal cords, which vibrate 70 to 250 times a second to produce sounds, are not the only things that distinguish your voice from that of someone else. The sound spectrograph shows that the size and shape of a person's vocal tract enhance certain frequencies, and the resonances of these frequencies make up the vocal signature. Such resonances, called *formants* by speech scientists, tell your listener's brain—or the spectrograph—that it is you and not somebody else who is speaking.

Even if you try to disguise your voice, the sound spectrograph is so sensitive that it will reflect the disguise. Professional mimics like Tim Russell or Rich Little, who can sound remarkably like the person they are mimicking, display voiceprints that are quite different from the person whose voice is being mimicked.

A sound spectrograph is used to identify voiceprints. The machine records sound frequencies and then converts them into signals that can be displayed visually.

Above: Lt. Lonnie Smrkovski of the Michigan State Police Voice Identification Unit. Right: Ira with mimic Tim Russell.

Most of us believe that we can tell something about a person's age by hearing his or her voice. Surprisingly enough, voice patterns do not change very much over time, so your voiceprint will remain pretty constant throughout your life.

Interest in this field began during World War II, when the U.S. military needed a reliable way to identify a speaker's voice. They enlisted the help of Bell Laboratories in Murray Hill, New Jersey to invent a mechanism for voice identification. No sooner was the device completed than it was called into use for analyzing scrambled voice patterns, studying voice-coding devices, extracting speech from noise and training the deaf.

In the late 1950s the New York City Police Department was being plagued with bomb threats to airplanes, and police investigators began to use the device. By 1968 both aural and spectrographic analysis were put to work in police investigation. The Michigan State Police Voice Identification Unit, commanded by our guest Lieutenant Lonnie Smrkovski, has examined thousands of voices in the last fifteen years. In one study of 192 speakers of general American English made under everyday conditions, decisions were made in 73 percent of the cases, and all decisions were found to be accurate. (In some cases, there is so much interference from background noises or technical problems that it is not possible to make a decision about a match between voiceprints.)

Because it is so effective in sorting out formants, the sound spectrograph is now used in legal cases that require positive identification. Just as with all expert testimony, this naturally raises questions about its reliability and accuracy. How are voices, and voiceprints, affected by stress, illness or

Professional mimics have voiceprints quite different from those they imitate

fatigue? We know that alcohol and drugs can both affect a voiceprint. So can depression. So can simpler adjustments to surroundings such as waking up, shouting in a crowded room or whispering. Quirks and variables in microphones and tape recorders can also cause distortions. Since voiceprints must be interpreted, they provide a sophisticated, but somewhat controversial, form of expert testimony. However, at least four major studies have shown that well-trained examiners can make highly reliable decisions.

Aside from these questions—and the ones over violations of the Fourth Amendment's protection against invasion of privacy—the first real test of admissibility as legal evidence took place under rather bizarre circumstances. A woman named Constance Trimble telephoned the St. Paul Police Department and reported that her sister was about to give birth and that she needed a ride to the hospital. When a responding officer knocked on the door he was fatally shot in the back from a high-rise apartment across the street. Trimble was charged with the murder, after which ensued a long legal wrangle over the admissibility of voiceprints as identification. After more than a year of debate, Trimble finally admitted in court that she had made the call. Since that time, hundreds of trials in the US have involved voice identification.

Why Do We Get GOOSE BUMPS?

Ira: Did you ever wonder why you get goose bumps when you are cold? Here to help us with the answer is Jan Serie. Jan, you usually bring Dead Earnest with you. Where is he? I hope he's not feeling under the weather.
Jan: I brought another one of my friends to demonstrate why you get goose bumps when you are cold.
Ira: Why a gorilla? I get goose bumps just looking at him.
Jan: We're going to talk about hair, and since Dead Earnest doesn't have any, this hairy guy volunteered to come on the show.
Ira: I don't want to sound inhospitable, but I hope this is going to be his last visit, although he somehow looks familiar.
Jan: I'm sure you have a lot in common. For one thing you both have hair. And when you and this gorilla are cold, you fluff up your hair or fur so you can trap warm air near your skin.
Ira: Like a bird does when it fluffs its feathers. But if I fluffed up my hair, nothing would happen. It is certainly not going to make me warmer.
Jan: No, no, it's not going to make you any warmer, but you have the same number of hair follicles per square inch as your friend here.
Ira: I wouldn't say he's my friend—we've just met.
Jan: According to the theory of evolution, when your ancestors looked like him, they fluffed their fur to stay warm. You still have that mechanism, even though it doesn't do you any good.
Ira: How do the hair follicles get up like that?
Jan: Each hair follicle is attached to a muscle, and when the sympathetic nervous system wants to warm you up, it makes the muscle contract. The contraction hoists the hair up in the air, which dimples the skin and creates goose bumps.
Ira: So I must have millions of hair muscles all over. It's nice to know I'm so muscular.
Jan: Each tiny hair has one.
Ira: What about the times when your hair stands on end because you're afraid?
Jan: The sympathetic nervous system again responds when you're afraid of something. The response is an adaptive one because if something big and mean and ugly were chasing you down a dark alley, you want to appear as big and mean yourself. You fluff up your hair to make yourself look larger.
Ira: Like cats and dogs do?
Jan: Exactly.
Ira: Thank you, Jan, for coming here and explaining that, and thank you also for bringing King Kong Jr. with you.

There is nothing very complicated about how a sound spectrograph works. It simply converts sound frequencies into visual displays. The machine does this by first recording the sound frequencies and then converting the magnetically recorded sound into electrical signals. An electronic stylus, similar to the ones that record earthquakes, transfers the electrical signals to a visual display by burning the surface of paper that is sensitive to electrical conduction. The result is a spectrogram that presents the acoustic patterns of an individual's voice.

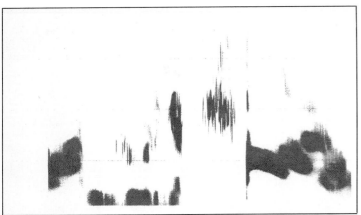

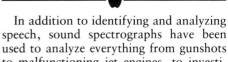

Analyzing gunshots and malfunctioning jet engines

In addition to identifying and analyzing speech, sound spectrographs have been used to analyze everything from gunshots to malfunctioning jet engines, to investigate noise, to improve soundproofing, to provide better communications equipment, and to identify aircraft, ships and submarines.

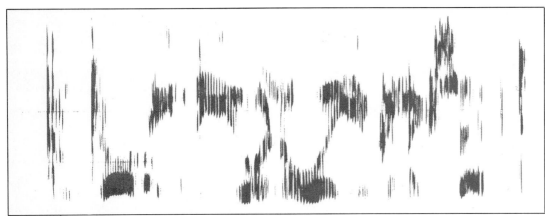

Tim Russell tries to fool the sound spectrograph with his imitations of Julia Childs and George Burns. Though his imitations sounded convincing, his voiceprints did not look anything like those of Childs or Burns.

Minnesota Zoological Garden

Red-tailed Hawks

Three hundred years before Christ the Chinese began training hawks to hunt, and thus was born the sport of falconry. The word *falcon* refers to any hawk trained by humans to hunt. By the fourth century A.D. falconry had spread to Japan. Dogs and trained hawks were housed close together in royal estates in the Orient, and they were so highly prized that taxes from outlying provinces were often paid in the form of hawks.

Falconry soon spread to Europe. Trading caravans from the Near East carried falconers with trained birds to Central Europe, where the sport became part of the culture. Later, Crusaders returned home having acquired the custom, and triggered an amazing preoccupation with hawking throughout Europe. Skill in handling falcons was considered an essential part of a gentleman's education, and the skill achieved a status and following never before seen in the western world.

Elizabethan England saw the flowering of falconry in the British Isles. Many subtle nuances in Shakespeare's plays are lost to modern audiences unfamiliar with the meanings and connotations of the hawking vocabulary. With the development of firearms more than 250 years ago, falconry lost its popularity as a means of hunting.

Taxes were paid in the form of hawks

One variety of this noble breed often seen in North America is the red-tailed hawk, for they are widely distributed throughout the region. Wild red-tailed hawks are fast and aggressive, and human-trained red-tailed hawks are gratifying performers in the field.

Hawks soar for hours over their territories, and in the spring they tend to stay a little closer to home while feeding their young. Their distinctive calls can be recognized when the birds themselves are barely visible in the sky. Red-tailed hawks are able to capture such elusive quarry as rabbits, squirrels and mice. On very rare occasions they have been known to prey on ducks, pheasants, starlings, and sometimes even house sparrows.

Red-tailed hawks may choose an inconspicuous site along the edge of a large forest, but they are usually found in much

more exposed situations. They frequently perch on telephone poles. In California, a red-tail soaring over a clump of eucalyptus in early spring almost certainly means that its stick nest is in the crown of one of the trees. Red-tailed hawks engage in spectacular display flights associated with nesting. These flights are readily observable and generally end up in the grove that holds the nest. Red-tailed hawks even nest in urban settings like parks.

Spectacular display flights at nesting time

Newton's Apple asked Nancy Gibson, our friend from the Minnesota Zoo, to visit us with a red-tailed hawk so we could find out more about it.
Ira: Nancy, why are these called red-tailed hawks?
Nancy: Like many birds, they are named after one of their distinguishable characteristics. After red-tailed hawks go through their first moulting period they grow a beautiful reddish brown tail, which is a sign that the bird is mature.
Ira: Is it a he or a she?
Nancy: It's a she.
Ira: And how can you tell?

they have been clocked up to 120 miles an hour. Their average hunting speeds are 60 to 80 miles per hour. As a hawk approaches its prey it extends its enormously powerful feet in front of it and squeezes the prey to death. Then the hawk spreads its wings over the prey, an action called *mantling*, so other hawks or predators can't see what it's eating. If the prey is large, the hawk will eat it on the ground. If it is smaller it will fly up into tree branches and eat it there.

Ira: Is this an endangered bird? One of the endangered species?

Not even a feather without a federal permit

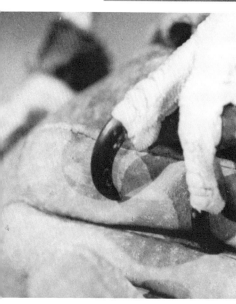

Nancy: It's very easy with birds of prey—eagles, owls, hawks, and falcons. The females are about a third larger than the males.

Ira: What is the hawk's greatest trait?

Nancy: It eats about one thousand rodents a year, which is very beneficial. If it didn't, we would all be walking around on stilts to avoid the prolific rodents.

Ira: We say that people have eyes like a hawk. What is so special about a hawk's eyes?

Nancy: Their eyesight is the most phenomenal hunting tool! Imagine that you are sitting on an airplane cruising five or six miles above the earth and you recognize one of your friends down below.

Ira: These birds can see a few miles in distance? How?

Nancy: At the back of their retina they have two small depressions called *foveae*. Unlike humans, who only have one fovea, hawks have two. This is the area of sharpest vision because of the high concentration of cones—ten times the amount in humans—that are responsible for daytime visual acuity. One fovea is for search and the other for pursuit.

Ira: And what happens when a hawk goes after its prey?

Nancy: When they're really in hot pursuit

Nancy: Some birds of prey are endangered, but red-tailed hawks are not. All birds of prey are protected, however, which means you can't even own a feather without a federal permit. They are protected because wildlife experts realize how important they are in controlling the population of rodents.

RIGHT MOVES

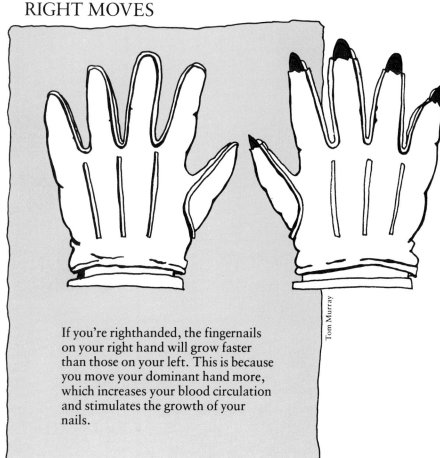

If you're righthanded, the fingernails on your right hand will grow faster than those on your left. This is because you move your dominant hand more, which increases your blood circulation and stimulates the growth of your nails.

Tom Murray

NEWTON'S APPLE ■ 25

Behind the Scenes 3

NEWTON'S APPLE

■ To demonstrate the keen eyesight of hawks, we planned to have Nancy Gibson of the Minnesota Zoo hold some food in her hand while Ira held the red-tailed hawk. We wanted the hawk to fly from Ira to Nancy and it worked every time we rehearsed it. ■ When we were taping in front of the studio audience, Ira released the hawk and it flew right up to the ceiling. There it perched on the lights for thirty minutes! The audience loved it.

NASA

Learning to Deliver

Astronauts who work on the space shuttle are proud of saying "We deliver." They have every reason to be proud of their space accomplishments, not only because the training and talent required for their work is awesome, but also because they do it so brilliantly. This level of performance brings some unusual payoffs. In *The Right Stuff*, a book about the astronauts, Tom Wolfe describes the feelings of the early astronauts this way:

"From the very beginning this 'astronaut' business was just an unbelievable good deal. It was such a good deal that it seemed like tempting fate for an astronaut to call himself an astronaut, even though that was the official job description. You didn't even refer to the others as astronauts. You'd never say something such as 'I'll take that up with the other astronauts.' You'd say 'I'll take that up with the other fellows' or 'the other pilots.' Somehow calling yourself an 'astronaut' was like a combat ace going around describing his occupation as 'com-

THE SPACE SHUTTLE

Space shuttle crew (left to right): Robert L. Crippen, crew commander; Frederick H. Hauck, pilot; John M. Fabian and Dr. Sally K. Ride, mission specialists.

bat ace.' This thing was such an unbelievable good deal, it was as if 'astronaut' were an honorific, like 'champion' or 'superstar'."

In the great American tradition, today's space shuttle astronauts have begun to specialize, and now there are down-to-earth job titles by which they can describe themselves. Newton's Apple went to NASA's Johnson Space Center and visited with Dr. Jeffrey Hoffman, a "mission specialist." He is one of the 35 new astronauts selected from more than 8000 applicants in 1978. As a mission specialist Jeff is in charge of the shuttle cargo, and he conducts scientific experiments in space as well as assisting the pilots who fly the shuttle. Jeff is one of those folks who "deliver."

Ira: We're surrounded by all kinds of shuttle vehicles and machines. What are all these used for?
Jeff: These are simulators, trainers. They're built to look just like the shuttle on the inside. We spend a lot of time in there, learning how to carry out tasks that we will have to do in space. Let's go up to the flight deck. Now, you should realize that if you were really getting into the shuttle before launch it would be vertical and we would be standing on our sides. So it would look very different from the way it looks now.

Ira: Our first stop is the orbiter trainer where astronauts learn how the space shuttle flies. During takeoff and landing, the two pilots sit in the cockpit. As a mission specialist, Jeff doesn't pilot the shuttle, but he is required to know his way around the cockpit, which has thousands of buttons, knobs and dials. Jeff, this panel here with three screens and a pushbutton board looks like it belongs to a computer.
Jeff: It does, Ira. We are heavily dependent on computers in learning to operate the shuttle. In fact, we can put our hands on the

From 19,000 miles an hour to a couple of hundred when preparing to land the shuttle

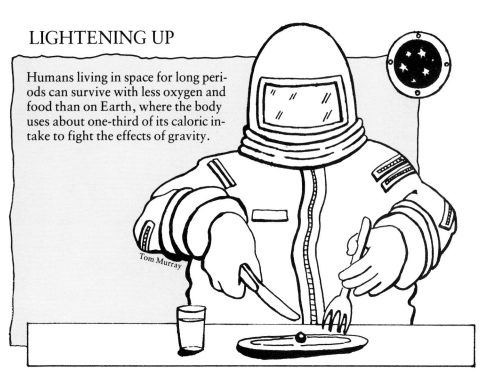

LIGHTENING UP

Humans living in space for long periods can survive with less oxygen and food than on Earth, where the body uses about one-third of its caloric intake to fight the effects of gravity.

controls to fly the shuttle, but what we're really doing is flying computers. We tell the computers what we want them to do and then they decide how to fly the shuttle. For example, when the shuttle returns to the Earth's atmosphere, we are flying 25 times the speed of sound—about 19,000 miles an hour. So we have to slow all the way down to a landing speed of a couple of hundred miles per hour. The shuttle's flight characteristics vary tremendously during entry into the atmosphere, and if you had to control individually every single aerodynamic control, you simply could not do it. When we tell the shuttle computers that we want to make a right turn by turning the control stick to the right, the computer then uses its knowledge of navigation and aerodynamics to decide, "Ah, yes, we're flying twelve times the speed of sound; therefore, we have to fire this many jets and we have to move the craft in such and such a way." The computation might be completely different if we were flying three times the speed of sound.

Ira: The space shuttle is the first step in

making space accessible to everyone, not just highly trained astronauts. In the future, passengers may be riding along on a shuttle as easily as we fly on an airplane today. It is not too fantastic to think that by the turn of the century there will be Earth-like space colonies housing thousands of people (who could be mining ores on the moon, manufacturing perfect machine parts, pure alloys and drugs in the almost-zero gravity of space, or erecting solar power satellites that would beam solar energy back to Earth), and these people would take a space shuttle back to Earth for a vacation in Atlantic City. But right now it's not as simple as calling your travel agent. Passengers will have to take an 80-hour course in emergency procedures and learn how to work some of the equipment on board. Jeff, where are we now?

For future space passengers, an 80-hour course

Jeff: We're down on the mid deck. During the launch and entry phase, this is where we set up passenger chairs. The mid deck is also where much of the equipment and supplies are stored, with everything clearly labeled and listed. Patches of velcro are everywhere—on the wall, the ceiling, my pants. When you're floating around in space there's no such thing as putting something down. It has to be stuck somewhere or stored away.
Ira: There is stuff up here that I could hardly reach, like that pantry food up there.
Jeff: That's when you're standing on the floor. It's different when you're standing on the ceiling. You have to think of yourself in relation to zero gravity.
Ira: In zero gravity, what would happen if I tried to push that drawer closed?
Jeff: You would float backwards.
Ira: Such a simple task would be that difficult to do?

"It's different when you're standing on the ceiling"

Jeff: Using one hand to do things becomes very difficult. If I had to push this thing in and I were weightless, I would have to grab onto it with one hand and then grab on here with the other hand so I could pull myself in and keep my body restrained. Notice that underneath here is a place to put your feet, so I would go over and get a toe hold to keep my body in place. Then I could reach out and push it in with no trouble.

Things are not always more difficult in space. Climbing down to the mid deck is a breeze when you can simply glide through the doorway. And if you hate to exercise, zero gravity is just the thing for you. Anyone can do one-handed pushups in space.

Velcro patches on clothes and walls

Keeping your body in good condition is a problem in a weightless environment. Astronauts have to tie themselves down to an exercise treadmill for it to work properly. Another adjustment involves eating. Meals consist mostly of casseroles with lots of sauce on them. The sauce has to be sticky enough to adhere to the package and the spoon, so your dinner doesn't float away. There is one more difference about living in zero gravity—going to the bathroom. You would be surprised at how similar the space shuttle toilet is to one on an airplane. The

Continued on page 32

Down

The zero-gravity aircraft shown at the top of the page is used by astronauts and research scientists to simulate micro-gravity environments. In the photograph above we see astronaut Richard H. Truly practicing eating in a zero-gravity environment—not an easy task! The photograph on the opposite page shows a typical meal for astronauts in space: smoked turkey, mixed vegetables, strawberries, and cream of mushroom soup. To the left is a prototype for the 50-foot manipulator arm that is used in launching and retrieving payloads.

to Earth Training

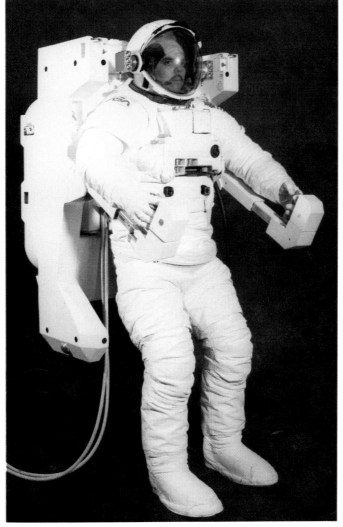

The space shuttle mockup and integration laboratory shown above is heavily used by planners and astronauts-in-training to prepare to carry out tasks in space. The photograph to the left shows the new "manned maneuvering unit" to be used in the January 1984 shuttle flight for repairing the Solar Max satellite. This unit is designed so that a shuttle crew member can fly with great precision in and around the shuttle's cargo bay or to nearby free-flying payloads such as satellites.

Continued from page 29

thing we lack up here is gravity to make everything go down the toilet, so we replace gravity with air flow. Fans inside create a suction that pulls everything down into the waste storage container.

Ira: It's also the only toilet I've ever seen that comes with its own seatbelt!

Jeff: You certainly wouldn't want to float away! Anyway, after I learned how to live in space, it was time to learn how to work in space. As a mission specialist, one of my responsibilities is supervising the payload bay area where satellites are stored until they are sent into space with the remote manipulator arm. Because in space the satellites are weightless, the mechanical arm does not have to be strong, just precise. To practice moving a weightless load, the astronauts use a huge helium-filled balloon. The operator stands at a control panel and manipulates the arm by moving hand controls that work much like the joysticks that control video games. Let's start with this arm.

Ira: I move this to the left and that should go to the left. And, sure enough, it's doing just what I asked it to.

Jeff: You're a success!

"I put on my pants two legs at a time"

Ira: I can see why you wanted to be an astronaut. This is fun.
Jeff: Oh, there are lots of neat things to do.
Ira: If I move back to the right, it goes right. It behaves like an extension of my arm.
Jeff: That's the way it was designed. After you have spent enough time using it, you come to know beforehand just what to do to place it in the position and direction you want. But the most fun is space walking. NASA calls it *extravehicular activity*, and I'm happy to say that is one of my jobs as mission specialist. For space walking you need a space suit, or what's called an *extravehicular mobility unit*.

This is the blue room where astronauts get fitted for their space suits. Each suit is a self-contained spacecraft, complete with a hard, fiberglass torso, ball-bearing joints, and a built-in computerized backpack equipped with oxygen and electricity. At the heart of its mechanical system is a 2-inch-diameter, one-twentieth-horsepower fan that circulates oxygen through the suit, removes carbon dioxide, and pumps cooling water through the tubular fibers of the astronauts' underwear. It doesn't look like it would be easy to put it on, but in space I can put my pants on two legs at a time! The rest of the suit is mounted to the wall. All I have to do is climb in. But it's your turn to put on a suit, Ira. Whoopsy daisy. There you go!

Weightless, you feel like Superman

Ira: It's a cinch. Let me at a moonwalk! I'm ready now.
Jeff: One difficult test involves a rotating chair. It is used to check out the astronaut's reaction to weightlessness, which sometimes brings on nausea. The idea is to examine an astronaut's threshold of sensitivity and then to prescribe appropriate motion-sickness drugs and repeat the tests to examine the effectiveness of the drugs. Space sickness has always been a problem for American and Russian astronauts. Half

Astronauts F. Story Musgrave, left, and Donald H. Peterson float about in the cargo bay of the Earth-orbiting space shuttle Challenger during extravehicular activity. Their floating is restricted via tethers to safety slide wires. Thanks to the tether/slide-wire combination, Peterson is able to move along the port side handrails. Musgrave is near the airborne support equipment. Clouds can be seen in the background.

of them have become sick. According to NASA flight surgeons, there is no way to predict which astronauts will become ill. As a consequence, mission directors now plan light work loads for the first day in space. Space sickness seems to be more of a problem on brief shuttle flights and less on the long missions lasting more than a month.

Ira: So they spin you around in this chair until you get nauseous and then they give you something for it to see if it works?
Jeff: All for the greater glory of science.
Ira: What does it feel like to be weightless?
Jeff: Like Superman.
Ira: Really?
Jeff: You can fly through the air, and everybody who has been up in space can't stop talking about that experience. It's just the pure delight of zero gravity.
Ira: Jeff has been showing me some of the more active and enjoyable parts of the training, but the complete astronaut training program is demanding and exhausting. Such training may last anywhere from five to fifteen years before a flight in space. But I have a feeling that not one of the astronauts would trade it for anything.

These beautiful cats are rarely seen because they dwell in the tops of trees, mainly in remote forests in Southeast Asia. Not only are clouded leopards hard to see, but taxonomists—the people who classify plants and animals—have more than one theory about how to classify these cats. Since they have the skull structure and the spots of a leopard, it appears they should be in the large cat family. However, the clouded leopard consumes its prey much like a small cat. To complete the confusion, its size puts it in the medium cat range. The taxonomists' solution was to put it in a genus by itself.

Ira: Nancy, how did this beautiful cat get its name?
Nancy: It has a cloudy-looking coat. It's not spotted in the way that most people associate with leopard spots.
Ira: How big do these cats get?
Nancy: The males weigh up to 45 pounds when they are full-grown.
Ira: That's not quite as big as those giant lions, tigers and leopards we're used to seeing.

Nancy: No, those grow very large. That's one reason clouded leopards are in a class by themselves.
Ira: I suppose it's difficult to research them in their natural habitat because it's so remote and they're so well-camouflaged in the trees.

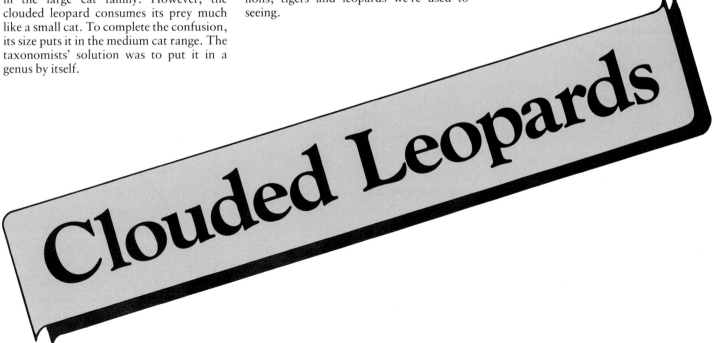

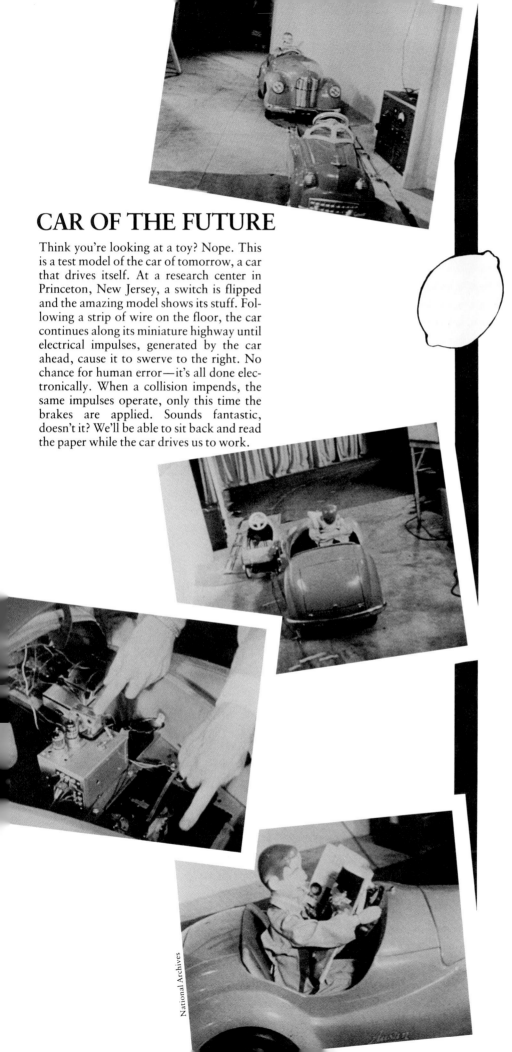

CAR OF THE FUTURE

Think you're looking at a toy? Nope. This is a test model of the car of tomorrow, a car that drives itself. At a research center in Princeton, New Jersey, a switch is flipped and the amazing model shows its stuff. Following a strip of wire on the floor, the car continues along its miniature highway until electrical impulses, generated by the car ahead, cause it to swerve to the right. No chance for human error—it's all done electronically. When a collision impends, the same impulses operate, only this time the brakes are applied. Sounds fantastic, doesn't it? We'll be able to sit back and read the paper while the car drives us to work.

Nancy: It's almost impossible to study them. Most of what we know about them is learned from the 150 clouded leopards in zoos around the world. We do not know how many are living in the wild, but they are an extremely endangered species because their habitats have been destroyed by wars and the spreading population of humans. Also, because they have such a beautiful pelt, they are killed to make coats.
Ira: He makes such nice purring noises.
Nancy: That's another way to classify cats. Generally, the large cats roar and the small cats purr, as you would expect.
Ira: OK, now let's see if we can answer one of the most frequently asked questions at Newton's Apple: What makes a cat purr?
Nancy: When small cats purr, they're exhaling and inhaling air through their voice box. The continuous airflow creates the

Habitats destroyed by wars and the spreading human population

Tom Cajacob/Minnesota Zoological Garden

purring sound. Purring is a form of communication, perhaps one of contentment. The larger cats roar because they have a more flexible voice box.
Ira: What do they eat when they get hungry, which I hope this one isn't?

Long tails for balance while climbing trees

Nancy: They eat small monkeys, rodents, small deer, some birds. They are well-adapted to living in trees and hunting their prey from treetops.
Ira: Does their long tail help them climb trees?
Nancy: It helps them to keep their balance. Plus they have large feet with very large claws.
Ira: He looks so cute I'd think some people would want a clouded leopard as a pet.
Nancy: He looks cute now, but once he's full-grown you would be risking your life if you tried to have him as a pet.
Ira: In that case I'll content myself with visiting him in the zoo.

CHASING THE COMMON COLD

Ira with Dr. Chuck Carlin

Medical researchers are no doubt sick of hearing the common expression of frustration "We can put a man on the moon but we can't cure the common cold." Yet who has not said it when suffering from a stuffy head and the other unpleasant manifestations of that all-too-frequent nuisance?

Our bodies contain about 100 million different kinds of proteins called *antibodies*. These antibodies help to kill the invading organisms that create disease, a process usually referred to as the *immune response*. With such an elaborate defense mechanism, it is puzzling that we keep catching colds or flu. The immune response to the common cold is precisely the same as to chickenpox or measles, yet few of us get through the year without a cold or a bout of flu.

Now research has shown that cold and flu viruses change and evolve from year to year, unlike those responsible for German measles or chickenpox. The antibodies that formed in response to previous colds provide only limited immunity, if any. So many different strains of colds and flu exist that immunity against one does not guarantee protection against another. And vaccines that are effective against one kind of flu can be totally ineffective in combating another strain of flu. This is because flu viruses can change their biochemical structure and create still another type of influenza.

The common cold is caused by a virus that infects the mucous membranes of the nose and throat, and not by exposure to wet or cold weather conditions. Even though most people believe that colds and flu, which are at their peak in winter, are caused by the inclement weather, it appears that this is not entirely the case. Colds are nonexistent in polar regions, where they occur only when introduced by visitors. In fact, common cold viruses cannot survive extreme temperatures. In moderate latitudes cold viruses exist in about equal numbers during summer and winter. Some people conjecture that the incidence of colds peaks in winter because we are confined indoors, breathing stale air in which viruses can accumulate.

Newton's Apple wanted to find out more about colds and their remedies, so we invited Dr. Chuck Carlin, a chemist and adviser to the Food and Drug Administration, to talk with us about colds and their remedies.

Ira: Tell us, Chuck, why we can go to the moon, as the saying goes, but we can't cure

Almost all cold and cough remedies contain the same chemicals

the common cold?
Chuck: There are too many different kinds of cold viruses. And they are all over the place.
Ira: How many cold viruses are there?
Chuck: More than 100.
Ira: Does that explain why no one has developed a vaccine?
Chuck: That's right. We can develop vac-

Ira: OK, let's see remedy number two. This is an interesting one. What have you got in the mug?
Volunteer: I've got a big mug of hot lemonade, a little honey, a pat of butter, and a shot or two of whiskey.
Ira: Ah, the secret ingredient!
Volunteer: Yes. And the butter makes it slide down easily. It tastes good. It's sweet. And it puts you to sleep so you rest and get rid of the cold.
Ira: At least you don't care about the cold anymore. Right, Chuck?
Chuck: That's right. It makes you sleep. If you feel lousy you might as well go to bed. And this will make you feel better and sleep better.
Ira: Some of the nighttime cold liquid remedies also have alcohol in them, don't they?
Chuck: Yes, and for the same purpose.
Ira: Here's our next remedy. Your name?
Momma: I'm Momma D.
Ira: Oh, you have a restaurant, don't you?
Momma: Yes, and every time anyone comes into my restaurant with a cold I get a spoonful of peppers, tell them to open their mouth, pop in the peppers and tell them to

---

The idea is to try to weather the storm

---

cines for polio or flu to help the body defend itself against these attacks, but with colds there are just too many organisms.
Ira: So what do all these cold remedies do for you?
Chuck: They help to relieve the symptoms—runny nose or sore throat—but they don't cure the cold.
Ira: Don't they contain mostly the same ingredients, although they may have different names for them?
Chuck: Very much so. Most of the cold remedies have the same chemicals, and the same is true for cough remedies.
Ira: There weren't always drugstores and there weren't always places to buy cold remedies, so people had to come up with folk remedies. We're going to try out a few to see if the old-fashioned methods work. Joan, you have a sandwich.
Joan: Yes I do. The infamous onion sandwich. My grandmother used this remedy and I've found that it works wonders. Your head becomes unclogged and everything drains out. The other advantage is that if you've eaten one nobody will come near you for the next four days, so you won't get their cold viruses.
Ira: I see you have some garlic here. Do you put that in the sandwich?

Colds are non-existent in polar regions unless introduced by visitors

---

Joan: My grandmother says I should, but I can't force myself to do that.
Ira: Well, Chuck, what do you think?
Chuck: Joan has a marvelous decongestant here. You eat the onion, your eyes run, you blow your nose, and you breathe more freely. So it makes you feel better.

shut their mouth and breathe. Before you know it, the tears run down and their sinuses open. I give some people a glass of wine and tell them to take it home and heat it as you do a cup of tea. Then I say to take two aspirin with it, cover up, sweat, throw out the wet garments and put fresh ones on. And they're OK the next morning.
Ira: And it works?
Momma: You're darn right. Grandmothers know what to do.
Ira: Chuck, what do you think of Momma's method?
Chuck: This is like going nuclear, compared with the onion-sandwich approach. These amazing jalapeno peppers clean out your head very effectively. You really do feel better.
Ira: OK. I'm going to tell you about my favorite cold remedy, old-fashioned chicken soup. Now all you do is heat it up and eat it. What do you think, Chuck?
Chuck: Well, it's not a remedy, but it sure makes you feel better. And you should do anything that makes you feel better.
Ira: That seems to be true of all our cold remedies. The idea is to try to weather the storm until the cold runs its course. In the meantime, I hope all your colds are little ones.

NEWTON'S APPLE ■ 37

Behind the Scenes 4

NEWTON'S APPLE

■ Ordinarily the Newton's Apple crew takes almost any situation right in stride, but the visiting snakes brought some pretty strong reactions. Our floor manager, who is responsible for relaying information from the producer to those on-camera, is very much afraid of snakes. For him, shooting the segment was a real lesson in self-control. ■ It was great to see the kids in the audience approach the snakes with curiosity rather than fear. Most of the rest of us let out big sighs of relief when Nancy Gibson took her wriggly friends back to the zoo.

HYPNOSIS: YOUR ATTENTION PLEASE!

Endorsements of hypnosis as a therapeutic technique by the American Medical Association and the American Dental Association have freed it from the stigma of questionable stage-show entertainment, but there is still an element of mystery attached to it in the public mind. Few topics have attracted as much speculation and misinformation as hypnosis, with more than 10,000 articles and over 1000 books published on the subject.

Not a trance, but a vivid state of concentrated awareness

The term *hypnosis* is misleading, as it comes from the Greek word meaning "sleep," and the hypnotic state is best described, not as a trance, but as a vivid state of concentrated awareness. It is actually an altered state of consciousness wherein the subject is deeply relaxed while highly focused on deliberate, attentive listening.

One of the earliest experimenters with hypnotism was Franz Mesmer, an Austrian physician of the eighteenth century. Mesmer found that a simple passing of hands over his patient would relieve and presumably cure the patient of what he called a "disharmony of the nerves." Until the middle of the nineteenth century, almost all practitioners of "mesmerism" followed Mesmer in attributing the effects they produced in their patients to the passage from the operator to the patient of some subtle physical influence or fluid that Mesmer had called "animal magnetism."

A disciple of Mesmer was Dr. James Braid, a Scottish surgeon who gave the name *hypnosis* to his subjects' "nervous sleep." Braid believed the state was not induced by animal magnetism, but was a suspension of the conscious mind. He felt that hypnotism could make the conscious mind unresponsive to pain, and therefore would be useful for inducing anesthesia in surgical patients. Although this proved to be true for selected patients, the introduction of chemical anesthesia eclipsed hypnosis in the surgical suite, although it is still used effectively for pain by many medical practitioners.

Sigmund Freud and his followers used hypnosis to relieve anxiety, heighten recall, and probe more deeply into the minds of their patients. At first Freud was intrigued by the process, but he began to think that its cures were only temporary and its uses limited, so he later rejected it. When Freud abandoned hypnotism in favor of psychoanalysis, the therapeutic world followed. By 1910 scientific and medical interest in hypnotism had faded.

A legitimate "therapeutic agent"—American Medical Association

After nearly a half-century of neglect, hypnosis once again came to the attention of therapists searching for better ways to help their patients, and in the 1950s both the American Medical and Dental Associations announced that hypnosis is a legitimate "therapeutic agent" for treating a variety of problems: backaches, insomnia, shyness, fatigue, overweight, asthma, drug dependence, mental depression, and high blood pressure, to name a few. In 1957 the American Society of Clinical Hypnosis was formed, largely as a result of organizing efforts by Dr. Milton Erickson, a pioneer psychiatrist-hypnotist.

The goal of the hypnotherapist is to lead the patient into a such a deep state of relaxation that it is possible to focus on material normally unavailable to the conscious mind. In most cases it is relatively easy to place a willing subject under hypnosis. There is no need for swinging a watch on a chain or producing a bright object to fix attention, but the patient must be physically and psychologically comfortable. Willingness and trust are key ingredients, and belligerence or a "see if you can do it" attitude are not conducive to hypnosis.

No need to swing a watch on a chain

People bring with them many misconceptions about hypnosis, so the first step must be to explain it and to eliminate the myths that have grown up about it. From the beginning of the session, conversation is guided in such a way as to reduce anxiety and increase receptivity to the hypnotic process. When the preliminary discussion is completed, the patient is asked to embark on a guided journey for which the therapist will be the navigator. It is essential that patient and therapist agree in advance about areas to be explored.

When the patient is ready, the hypnotherapist begins to speak in a calm, rather monotonous voice, encouraging the subject to lay aside any concerns of the moment and to concentrate exclusively on the process. It is suggested that the natural desire to evaluate what is happening be put aside until the session is over. When it is clear that the subject is completely relaxed, the hypnotherapist gently leads into a discussion of the desired topic. At the end of the session, the patient is slowly led back to a fully conscious state through the words of the therapist.

Responses to hypnosis vary from one individual to another, but some sensations are quite common. Often-noted sensations include heaviness or lightness, expansion, enhanced alertness, numbness, rising or falling, and time distortion. Hypnosis is different for everyone and there is no single phenomenon to describe it. A common reaction following hypnosis is that "nothing happened." This seems to stem from the misconception that the hypnotized person will be unaware of surroundings during the process and have no recall after it is over. Only at the deepest levels is this true, yet much useful work can be done without losing awareness of the outer world.

Many people have learned techniques of self-hypnosis to help them deal with stress and other problems of daily life. The process is the same except that it is self-directed. It begins with deep relaxation and continues with the focusing on a topic or problem of interest. Autosuggestion, as this is sometimes called, simply means that a person in a state of high concentration tells himself how to deal positively with a physical or psychological situation. This technique can be learned from books, but many people prefer to learn it from a professional.

Hypnosis is different for everyone and there is no single phenomenon to describe it

In a hypnotic state the subject is wide awake and focused within the self. A hypnotized person is capable of perceiving, evaluating, reasoning, deciding and carrying out his or her own decisions. No longer the province of charlatans, hypnosis has come into its own as a simple and natural way of dealing with certain physical or psychological problems that arise in our daily lives.

A simple and natural way of dealing with certain problems

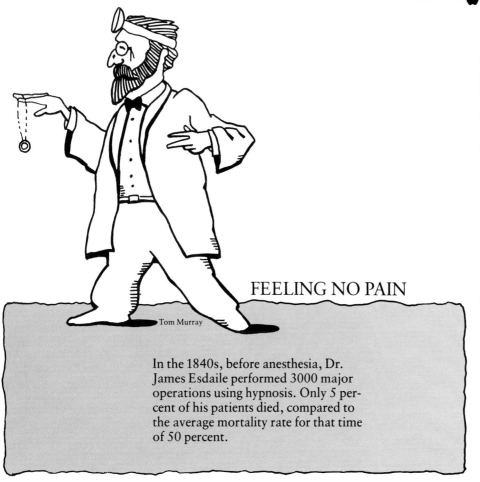

FEELING NO PAIN

Tom Murray

In the 1840s, before anesthesia, Dr. James Esdaile performed 3000 major operations using hypnosis. Only 5 percent of his patients died, compared to the average mortality rate for that time of 50 percent.

The material for this article appears courtesy of Glenn R. Williston, a hypnotherapist practicing in San Francisco, California.

A hypnotherapist demonstrates the powers of hypnosis with his subject. The woman is concentrating on being in Jamaica because she finds it very relaxing. As she and the therapist proceed with the demonstration, she visualizes a group of multicolored balloons that "allow" her arm to become lighter and lighter as she drifts into a more relaxed and enjoyable state. At the same time her arm becomes rigid to the point of being like a solid steel bar. (Ira tried in vain to bend her arm.) Hypnosis can be useful as an anesthetic in surgical procedures, and it is also valuable in teaching people how to control pain.

Snakes

If it wiggles like one and hisses like one, it is one; so the saying goes, but nature often contradicts these generalizations. There are approximately 6000 different types of reptiles in the world, and with all the different-looking animals that are called reptiles, a wiggle or a hiss is scarcely adequate to identify the members of this family that came to land in search of insects 250 million years ago.

A general characteristic of all reptiles is that they are cold-blooded animals. This does not mean they have cold blood. They

Snakes use their tongues to get their bearings

are *ectothermic*; that is, they regulate their body temperature within a certain range which is dependent on the external environment. This explains why lizards, snakes, turtles and crocodiles are often found sunning themselves. Other general characteristics are that the majority lay eggs, have a bony skeleton (and hence are called *vertebrates*), and breathe air. Their brains are poorly developed.

Snakes are often thought of as slimy, but

Tyrannosaurus rex (opposite page), now extinct, was one of the largest known reptiles. There are 18 orders of reptiles, but only four still exist: tortoises and turtles, lizards and snakes, crocodiles and Rynchocephalia. Snakes evolved from limbless lizards, and one way to tell the difference between them is by the length of the tail. The python (right) is only about 1/10 tail, while the glass lizard (below) is about 2/3 tail. Another difference is that lizards have eyelids and ear openings, while snakes have neither.

they are actually dry. Certain lizards, which are made up of two-thirds tail, drop their tails if they are grabbed by a predator. The lizard simply leaves it behind, thrashing the ground and distracting the predator while the rest of the lizard scurries away. Amazingly, the lizard grows another tail, although not as attractive as the original one.

Lizards and snakes differ from each other in several respects. First, snakes have a shorter tail in relation to their total body size than do lizards. A ten-foot python has a tail that is almost a foot long (males have a longer and heavier tail than females). Another obvious difference is that lizards have eyelids and ear openings. Some people think snakes are evil-looking because they lack eyelids. They also lack ears, but they do sense vibrations around their body, and that is what they use for "hearing."

A common myth about snakes is that when they stick out their tongues they are ready to strike. In fact, snakes use their tongues in place of the senses of smell and taste to help them find their bearings. Snakes keep flicking their tongues because they are sampling air particles, which are carried to their Jacobson's organ; this organ determines what the tongue was sensing. So it's important that a snake's tongue be in good working order.

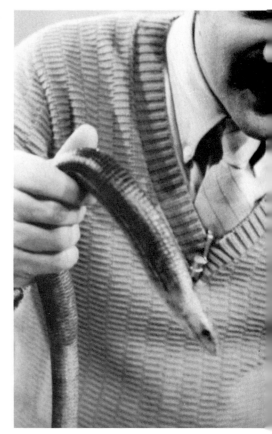

We invited Nancy Gibson from the Minnesota Zoo to come on the show and talk about snakes and other reptiles.
Ira: Nancy, you've got a pretty big python here. Is this as large as it will grow?
Nancy: No, its growth is determined by how much food it gets to eat. As it grows, it sheds its skin. The record length for a reticulated python was 33 feet. The anaconda can grow a little longer, up to 38 feet.
Ira: That's a mighty long animal. This one has been well-fed, I hope.
Nancy: Right now we're giving her four rats a month.

Reptiles are cold-blooded animals, but that doesn't mean they have cold blood

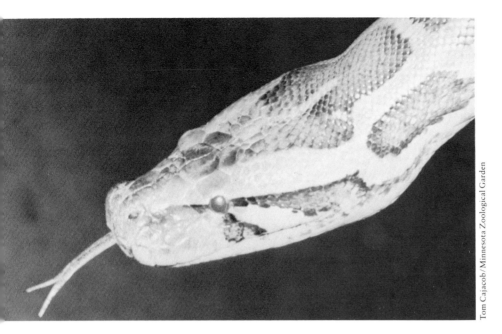

Ira: Don't they choke or suffocate when they are swallowing a large object?
Nancy: No, they can breathe through their nose, which also helps when they're swimming, because they can keep their noses above the water.
Ira: Is this pretty brown color for camouflage?
Nancy: Yes, and it's an ideal camouflage for the python. It sits on the ground and ambushes its prey. It grabs the prey with teeth that are slanted backwards, then the carnivorous python will wrap around its prey until it suffocates. Then it swallows the prey whole, head first.
Ira: I sure hope that it doesn't get hungry before the show's over.

MATCHLESS CIGARETTES

Forget to fill your lighter or bring a match? Well, things are tough all over. But here are a couple of nicotine devotees who have a new wrinkle. These new-fangled cigarettes strike and light right on the package. Where there's smoke, there must be fire. And where there's fire, yup, that's a smoke! The new style in self-lighting cigarettes was sparked in Italy. The striker is attached to the package. So strike up, sit back and enjoy yourself, because with this bright idea you can sure save yourself a pack of trouble.

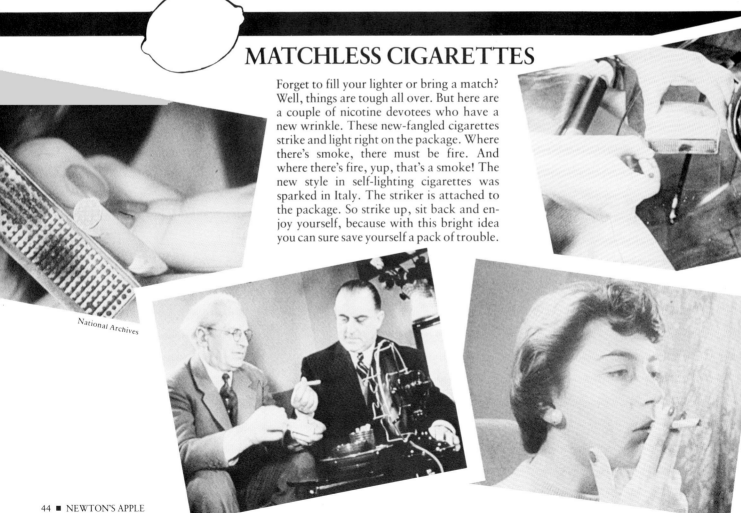

WHY IS THE SKY BLUE?

GREEN

YELLOW

RED

A classic question of childhood asks why the sky is blue, and many a bewildered adult has finally resorted to "just because it is." Newton's Apple to the rescue!

The sky looks blue because atoms, molecules and tiny particles in the sky selectively scatter light waves. The tinier the particle, the higher the frequency of light it will scatter. You can compare this to the ringing of bells. Small bells ring at higher sound frequencies (higher notes) than large bells do. Nitrogen and oxygen molecules that proliferate in our atmosphere are analogous to the small bells in that they are made to ring at predominantly higher frequencies.

Particles selectively scatter light waves

The ultraviolet light of the sun is scattered the most by the nitrogen and oxygen molecules in our atmosphere, although much of the ultraviolet light is absorbed by the ozone gas of the upper atmosphere. For visible light, violet is scattered the most. Similarly, some blue light and even less green light is scattered in the same way. Red light is scattered only about one-tenth as much as violet light.

Our eyes more sensitive to blue than to violet light

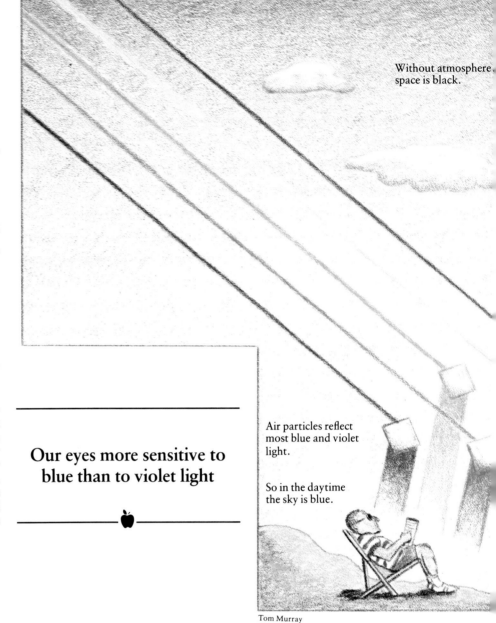

Without atmosphere space is black.

Air particles reflect most blue and violet light.

So in the daytime the sky is blue.

Tom Murray

When sunlight enters the earth's atmosphere, violet and blue light are scattered the most, followed by green, yellow, orange and red, in that order. For every ten violet photons (or discrete packages of light) that are scattered, only one red photon is scattered. This means that the sunlight scattered by the atmosphere has a predominantly higher frequency. Even though more violet photons are scattered than blue ones, our eyes are not as sensitive to voilet light as they are to blue light. And that's why we see blue skies.

When the air is filled with particles of dust and other particles larger than nitrogen and oxygen molecules, the lower frequencies of light are scattered more and the sky seems less blue; it takes on more of a whitish appearance. But after a heavy rainstorm, when the particles have been washed away, the sky becomes a deeper blue.

The lower frequencies of light are scattered the least by nitrogen and oxygen. So

RED SKIES

Mars really is the "Red Planet." Even its atmosphere is filled with reddish-brown dust particles blown into the sky by wind storms on the Martian surface.

Tom Murray

At sunset, direct light has lost its blue in distant skies, so the sun looks red.

red, orange and yellow light are transmitted more readily than blue light. Red light, which is scattered the least, travels through

At day's end, the sun's light travels further through the atmosphere

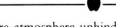

more atmosphere unhindered by nitrogen and oxygen particles than do any other colors.

At the end of the day, the path of the sun's light is at an angle to the Earth's surface that causes it to travel a greater distance through our atmosphere. When light passes through this thicker atmosphere during sunsets, the

After a cleansing rain, a deeper blue

higher frequencies are scattered, while the lower frequencies are transmitted. This explains why the yellow-white sun appears to be red at sunset.

At noon the sun appears to be yellow, because the path of sunlight through our atmosphere is most direct and therefore at its shortest, so a smaller amount of blue is scattered from the white light. (When blue light is subtracted from white light, the complementary color that is left is yellow.) The sun then becomes progressively redder, going from a yellow to an orange, and finally to a deep red-orange just before setting.

Because atmospheric conditions vary from day to day, we are lucky enough to have a glorious variety of colors in sunsets.

Why Do We See STARS When We Are Hit on the Head?

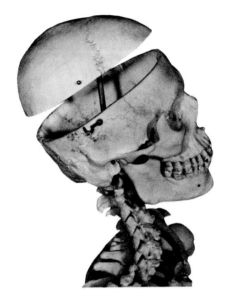

Ira: Why do you see stars when you're hit on the head? An interesting question that we hope one of our favorite stars, Jan Serie, can answer. Good to see you, Jan. And Dead Earnest looks in good form.
Jan: He's rather scared about being hit on the head, but we're going to demonstrate why you see stars. Although the mechanism is not very well understood, we do know that a minor blow may cause a random discharge of electrical impulses from your retina. Within the cranial cavity, as it is called, your brain is protected by a thick, bony cover. Sometimes, though, if you get hit really hard, the force is transmitted into the brain and you get what is called a *concussion*. The part of the brain that you use to see is located in the back of the head; it's called the visual cortex. When you get hit, the force is transmitted to the visual cortex and the nerve cells there discharge, causing you to see flashes of light. Actually, stars may be caused by electrical discharges at any point along the optic nerve. What can also happen is that the auditory center may randomly discharge, causing you to hear a ringing in your ears, or the sensory center may discharge, causing you to feel a tingling in your arms.
Ira: Just taking a New York City cab can create all those things in me. Anyway, it's shock waves transmitted through the skull?
Jan: That's what some scientists think.
Ira: But it's something that is transitory. Not a lasting effect, right?
Jan: Not under normal circumstances. But if you get hit hard, some swelling can kill brain cells. Then you have serious damage.
Ira: Is that what happens to boxers who are punched in the head all the time? Don't they sometimes lose their ability to speak?
Jan: Yes. Boxers can sustain permanent injury from swelling in the brain. It can cause lifetime disability.
Ira: The problem, then, is to prevent the swelling?
Jan: If someone is hit on the head and afterward goes to the hospital, doctors watch for signs of swelling and try to prevent it. Once the swelling has occurred, though, permanent damage may result.
Ira: Well, here's hoping that the only stars we see are in the sky or at the movies.

Behind the Scenes 5

NEWTON'S APPLE

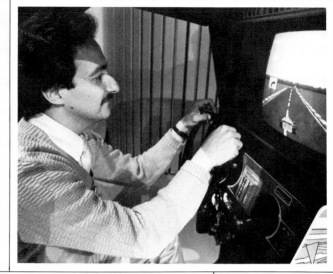

■ Video game fever struck on Newton's Apple with just as much force as it has hit the rest of the country. It was virtually impossible to tear Ira and the crew members away from the video game so we could get on with the segment.

VIDEO GAME FEVER

Americans love their fads, and video games have captured our minds, eyes, and quarters to become one of the most phenomenal fads of all time. Countless hours are devoted to playing them, and more money is spent on video games than on all other kinds of entertainment. All those quarters plunked down the slots of video games add up to billions of dollars a year.

What goes on inside these machines is as intriguing a question as what goes on inside the minds of those who make and play them. Video games are controlled by tiny computers on silicon chips. These microprocessors, as they are called, are the video games' brains. Crammed with enormous amounts of information, they tell the hardware—the things you can put your fingers on—what sights and sounds the video game should make.

Ira: We have with us Owen Rubin, a programmer at Atari, the makers and shakers of video games. Owen, how do you design a game like Millipede here so that it can show all these spiders, inchworms and earwigs moving on the screen?
Owen: The pictures are drawn on a video screen, in just about the same way that Saturday morning television cartoons are drawn. Someone draws a lot of pictures of these creatures, and then the computer rapidly displays them one after another to create the illusion of animation. At the same time the computer animates the objects, they are moved across the screen. So you have animation along with movement.
Ira: And all these pictures are stored in the hardware?
Owen: That's right. The pictures change 30 times a second, which is similar to how television images are displayed.
Ira: Now when this arrow on the screen is shot at a dragonfly, for instance, does the computer remember it?
Owen: Yes. Every time something moves, the computer compares the move with the

Guest Owen Rubin of Atari introduces Ira and the audience to the various components of a video game.

position of other objects on the screen to see if the arrow is going to hit something.

Ira: How does the computer know when something is hit?

Owen: It knows the horizontal and vertical position of the object, and it knows that the arrow was at the exact same place. This sets off a response from the computer program something like "Oh, he just hit that. Well, then, we need to show an explosion." With other objects that are not designed to explode, like the millipede, imagine the computer program saying "Since the millipede can't hit it and blow it up, I'll just turn him out of the way."

Ira: So the program tells the millipedes to distinguish friendly objects from hostile ones. What about these dragonflies and spiders and what-not that fly down from the top of the screen? It looks as though their motion is random.

Owen: It is. A special device generates random numbers that tell the spider and others when to move across the screen. Some games, especially older games, like Pac-Man, operate strictly with patterned movement—the same thing over and over again. When randomness is added, you can't second-guess the machine and so you have to improve your reflexes to try to beat the game.

Ira: How does the machine know what I want it to do?

Owen: It reads messages sent by a device called a track ball. Inside the machine all messages received by the track ball are translated into binary code (the numbers 1 and 0). When you move the track ball, these messages tell objects on the screen how

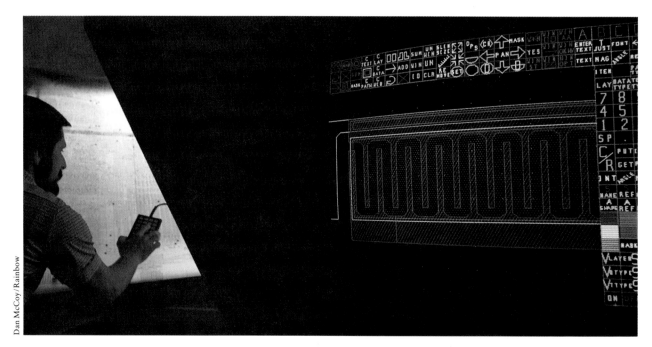

Circuits for video games are designed with the aid of a computer, as above, and then checked over manually at a scale thousands of times the final size of the finished product, as you can see by comparing the size of the circuit diagram at the left with the chip held by an ant on page 52.

Secret to winning: Practice, practice, practice—and lots of quarters

Ira: What's the secret to winning a game like this?
Owen: Practice, practice, practice. And lots of quarters!

The creation and production of video games is complex, fun, aggravating and painstaking. Ideas for games are developed in group brainstorming sessions and have come from computer whizzes still in their teens.

After an idea is agreed upon, a team of computer programmers, hardware design engineers, technicians and project engineers spend anywhere from six months to two years designing the game. A programmer writes the software for the game, and then a hardware engineer adapts the microprocessor to read the instructions the programmer has written. Once the game has been designed, it is tried out on something called a "breadboard," which consists of printed circuit boards that contain integrated circuits. With the help of an interactive graphics computer, the gamemakers design the integrated circuits so they fit onto the silicon chips. It's an awesome conversion of volumes of information onto a chip the size of your thumbnail.

When the game's internal circuitry is completed, graphic designers begin on the cabinet and controls. Ideas are formulated and then mocked up to show how the game fits together. At this stage the character or personality of the game comes into focus,

asks the players questions about the game to see what refinements need to be made. Then back they go to the design labs to redesign and rework sections of the game. Testing continues until the game is as good as it can be. After that, it is outfitted with its cabinet, boxed and prepared for shipment to the eager hordes of players.

No clearcut formula exists for creating a successful game. The PacMans of the business become "star" games only once in a great while. When you consider that the average price of a new video game is $3000, and the average lifespan is 12 to 14 weeks, the business has higher risks than one might assume (although the profits on a popular

Designers sit behind one-way mirrors to watch the test players' reactions

and the objects and their actions take on two-dimensional form.

Then comes the job of testing the game. A group of "real players" usually plays the game's prototype while the designers sit behind a one-way mirror to watch the players' reactions. They observe while a moderator

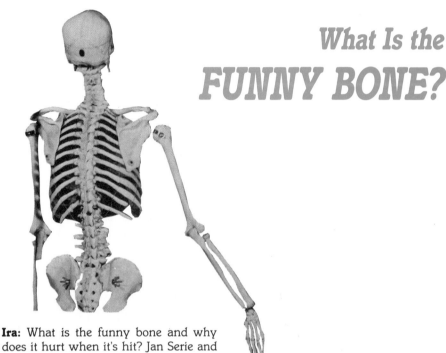

What Is the FUNNY BONE?

Ira: What is the funny bone and why does it hurt when it's hit? Jan Serie and her bony friend, Dead Earnest, are here to tell us.
Jan: I brought Dead Earnest to show you that the funny bone is not a bone at all, but a nerve. We dressed up Dead Earnest with a green wire representing a nerve.
Ira: He always seemed a bit wired to me.
Jan: This nerve, called the *ulnar* nerve, originates in the neck and runs down the arm to the elbow. It goes behind this bony place called the *medial epicondyle* and continues on to the hand, where it provides sensation to the fingers and helps to move them. The nerve passes behind this bony place, and when you hit your elbow on something, it pinches the nerve behind the bone. And that hurts!
Ira: You can say that again.
Jan: The nerve sends an impulse to the brain to let you know you shouldn't do that.
Ira: So you sometimes feel that tingling in your fingers because the ulnar nerve runs down to your hand?
Jan: Yes. They'll tingle.
Ira: If we have so many bones and nerves, why don't we have more funny bones? Not that I want any more.
Jan: The body has many bones and nerves, but the nerves are usually embedded in soft tissue—in muscles or other places that absorb the shock. A funny bone has to be right behind a bone where it can be pinched and hurt.
Ira: There is no other place in the body where that can happen?
Jan: A few places, like behind your ankle.
Ira: My ankle? I never hit that one.
Jan: It's not easy to hit, and also strong ligaments in the ankle protect that joint. The most vulnerable place by far is the elbow. Almost everybody has hit the funny bone at some time or another.
Ira: And sometimes so badly that it hurts for a few days or more?
Jan: Yes. The area will swell. Anything that you hurt is going to swell, and then the swelling pushes the nerve against the bone. If you move it a certain way, it will hurt again and send another impulse to your brain, which will remind you that you shouldn't have done that.
Ira: It gives me the tingles just talking about it.

game are staggering). The games are marketed mainly through the traditional distributors of pinball machines, and are found in places ranging from grocery stores to video arcades.

Who plays these games? According to one survey by Atari, not an unbiased surveyor, over 80 percent of the people between the ages of 13 and 20 have played gameroom video games. Young people who have short attention spans in school are content to spend long hours perfecting their video game skills. Isaac Asimov, a guru of science fiction, explained the attraction of the game like this: "Kids like the computer because it plays back. You can play with it,

---

**It doesn't say
"I won't play"**

◆

---

but it is completely under your control, it's a pal, a friend, but it doesn't get mad, it doesn't say 'I won't play,' and it doesn't break the rules. What kid wouldn't want that?"

It is generally acknowledged that manual dexterity may be enhanced by playing video games, and prospects for using them to teach difficult concepts like theoretical physics are encouraging. Many social psychologists and educators are convinced that video games teach players how to think in spatial terms better than conventional methods. Video games' use of *hyperspace*, which means that something can disappear from one reference frame and reappear in another, can help to develop a multiplicity of thinking patterns that were more difficult to teach before Nolan Bushnell came along, created Pong, and started Atari not much more than a decade ago.

As fads change, so will the nature of video games. Many designers are creating more educationally oriented systems, such as those that can teach people to animate cartoons, compose music, control robots, develop intuitive faculties, learn computer languages and, in general, enrich their minds. As computers become more adaptable to their users, and as the games and simulations become three-dimensional (perhaps with the help of holography), who knows what magic will emerge from what some call "microworlds."

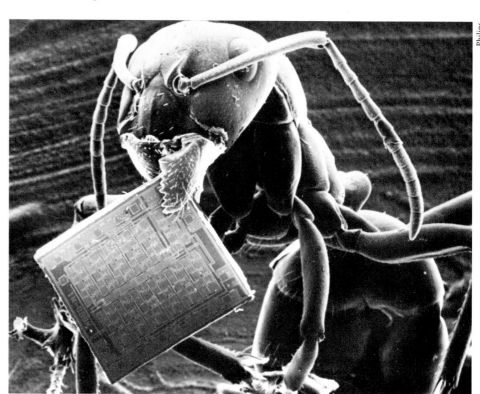

This photograph of an ant with a computer chip in its mouth shows the degree of miniaturization involved in computer chip design.

AUTOMOTIVE VIDEO

The coin slot in the first video game was a car radiator hose and the coin collection box was a one-gallon gas can.

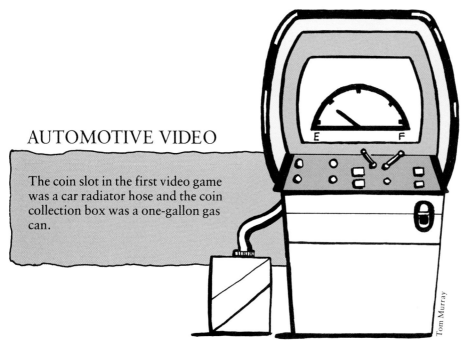

HOLOGRAPHY:
Taking Pictures in 3-D

Ira: I'm looking at a thin sheet of plastic surrounding apples that seem be mysteriously suspended in air. Are they really there? What is going on here? To tell us we invited Ron Erickson, educational advisor for the Museum of Holography in New York City, to explain what goes on in a hologram. Ron, what exactly are we looking at?

Ron: You're looking at light that bounced off those apples some time ago. The light was captured on film and frozen in time, and with special ways of reconstructing that light we can make the apples reappear just as they were when they were recorded.

Ira: I know that regular photography also captures light, but is this hologram capturing light in a different way? A regular photograph certainly doesn't look like this.

Ron: What is recorded in a hologram is different from what is recorded in a photograph, because in a hologram we record not only the brightness coming off an object but also the direction the light is traveling.

Ira: It's truly amazing. I see that there is no lens used in projecting this hologram. That's also totally different from photography, right?

Ron: That's correct. With a photograph, we have one and only one point of view, because all the light has to go through the lens in the camera to recreate that point of view. With holography, though, we've learned how to control light so we don't need a lens anymore.

Holography and photography: two ways of recording on film information about what we view. Yet how differently they accomplish their purpose and how different are the results. As the words *holo* (complete) and *gram* (message) connote, the hologram captures the entire message of the scene in three dimensions. A photograph records images in two dimensions and shows mainly brightness and, if

King Tut looks truly three-dimensional, yet in reality this disk is only an eighth of an inch thick. Viewed from an angle, the king still appears in three dimensions.

Some credit card companies are planning to convert their conventional plastic cards to hologram cards. Because holograms are so difficult to reproduce, they are considered a strong protection against forgery.

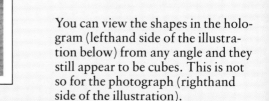

You can view the shapes in the hologram (lefthand side of the illustration below) from any angle and they still appear to be cubes. This is not so for the photograph (righthand side of the illustration).

color film is used, color. The hologram records the brightness of the light from the object, but it also records the precise direction the light is moving as it comes from the object.

The hologram plate resembles a window with the imaged scene appearing behind it in full depth. Viewers can look at the scene from many angles. To see around an object in the foreground, viewers simply raise their heads or move them from left to right. Contrast this with the old two-photograph stereopticon pictures that provided an excellent three-dimensional view—but still only one view.

To see around an object in a hologram viewers simply move their heads

All energy waves travel at a certain speed and wavelength (the distance from the crest of one wave to the crest of the next). Waves of light have the highest velocity of all. Light waves travel 186,000 miles a second, which is extremely fast compared to the sluggish pace of sound at 760 miles an hour. Because the hologram is a recording of light wave patterns, the length of the light wave plays a key role in hologram production.

When two light waves with the same wavelength meet, a phenomenon called *interference* results. That is, if the crest of one wave meets the trough of another, the wave is destroyed. Conversely, when the waves are superimposed on one crest, the effect of the wave increases. Color effects in soap bubbles are caused by the interference of light waves.

That light waves interfere with each other to form images has been known for some time. But it was not until 1947 that a British scientist, Dennis Gabor, invented an ingenious method for photographically recording a three-dimensional image. Even though Gabor conceived his idea fairly recently, in 1947 there was simply no light source known that could easily demonstrate the full potentials of holography. It wasn't until 1963 that an American scientist, Emmett Leith, linked the newly discovered laser with Gabor's technique for creating holograms. (The word *laser* is an acronym for Light Amplification by Stimulated Emission of Radiation.) Leith and his colleague George Stroke, the man who proposed the term *holography*, went on to make many important contributions to the field.

In forming a hologram, two beams of single-wavelength light waves are made to interfere with each other so as to record a three-dimensional image. One beam, called the *object beam*, focuses on the scene to be photographically recorded; the other beam, aimed directly at the photographic plate, is called the *reference beam*. When the reference light and the object light meet, they form an interference pattern on the emulsion of a photographic plate, where they record a pattern of fine lines. The pho-

Light waves interfere with each other to form holographic images

tographic plate, which is made of either a piece of film or a plate of glass coated with an emulsion, is then developed and fixed.

After photographic processing, the hologram will show its image only when you shine a light on it from the same direction as the reference beam. Then all the fine lines etched as a microscopic pattern into the emulsion reveal a faithful reconstruction of the scene's original dimension and depth.

The light needed to record the complicated pattern of a three-dimensional object must be highly directional. Light emitted by a common lamp is *incoherent,* which means it emits photons of many frequencies and many phases of vibration. The light is as incoherent as the footsteps on an auditorium floor when a mob of people are chaotically rushing about. A beam of incoherent light spreads out after a short distance, becoming wider and wider and less intense with increased distance. Incoherent light does not lend itself to holography. A laser, on the other hand, produces a beam of *coherent* light—one where the photons have the same frequency, phase and direction, like people marching precisely in rows. Only a beam of coherent light will not

Ira and Ron Erickson try a new method of bobbing for apples and come up empty-handed.

The word laser is an acronym for "Light Amplification by the Stimulated Emission of Radiation"

spread and diffuse, and is necessary to make a hologram.

Holograms come in two kinds: transmission and reflection. They are easy to identify. Transmission holograms have light shining through them. Some can be lit only with lasers. Others, called *rainbow holograms*, are lit with ordinary light bulbs. Rainbow holograms change color as you move up and down. Rainbow holograms and holographic movies can be lit with light bulbs because they break up white light from the bulbs into the seven colors of the spectrum. Such holograms act like prisms and present the image in each color as you move up or down.

Reflection holograms can be displayed with white light because the hologram filters out all but one color to form the image. That's why most reflection holograms

show an image in one color, although they can be made to form more than one image in two or three separate colors.

Holograms of stationary objects are the easiest to make because they require a relatively inexpensive laser that produces a continuous beam of light. Holograms of moving objects—people or mechanical objects—are more difficult to make because they require a pulsed laser, which is expensive and complicated to operate. Another, more popular, technique for making holograms of people or moving objects is called *holographic stereography,* which combines filmmaking techniques with holography. The subject is carefully filmed with a motion camera, recording the subject from many different points of view. Then all these flat images are combined holographically to produce a composite three-dimensional image that moves. The finished hologram is called a *holographic stereogram* or *holographic movie*.

An example of a three-dimensional holographic movie is "The Kiss" by Lloyd Cross, an important contributor to holography. As the viewer moves in an arc of about 120 degrees around the illuminated cylinder, the viewer sees the woman in the hologram first wink and then blow a kiss.

Although most holograms are made simply to be enjoyed, artists are turning holography into fine art and the commercially minded are creating myriad applications:

PIGEON PIX

Pigeons can't walk and see at the same time. They bob their heads because they can't adjust their focus while they waddle along. They stop moving their heads between steps to focus, much like a camera taking a series of snapshots.

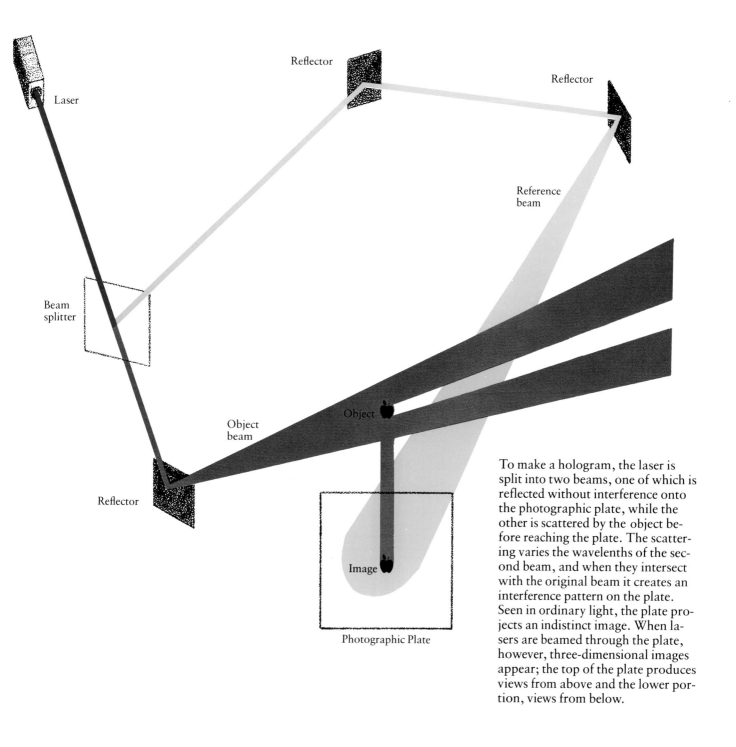

To make a hologram, the laser is split into two beams, one of which is reflected without interference onto the photographic plate, while the other is scattered by the object before reaching the plate. The scattering varies the wavelengths of the second beam, and when they intersect with the original beam it creates an interference pattern on the plate. Seen in ordinary light, the plate projects an indistinct image. When lasers are beamed through the plate, however, three-dimensional images appear; the top of the plate produces views from above and the lower portion, views from below.

advertising, jewelry, charge cards that foil forgers, video games, and a source for information storage. In industry, engineers use holography to make accurate measurements and to study the behavior of materials. A technique called *holographic interferometry* employs the detailed image in a hologram to make measurements with an accuracy of ten-millionths of an inch. Holographic interferometry is also used for nondestructive testing of products and parts, because it is possible through its use to "see" inside and determine if something is properly constructed. In the tire industry, for example, this holographic technique helps engineers to see if the bonds between the layers of tire are uniform or if they have gaps. In aircraft, specialized holographic lenses project instrument information out in front of the aircraft so the pilot can see the controls without looking away from the sky.

Holographic interferometry is used for testing products without destroying them

With the improvement of laser technology and holographic techniques, the capabilities and applications of this field are only limited by the imagination. Anyone for 3-D billboards?

IS THERE ANY WAY TO CUT ONIONS WITHOUT CRYING?

Whole onions are fragrant, but slice them and weep

Why is it that we break into tears when we slice through the seemingly harmless yellow bulb of an onion? A basket of onions sitting on the kitchen counter may be fragrant, but never a cause for tears. Clearly, the slicing is where our trouble begins.

Within the onion are innumerable cells whose membranes separate a sulfur compound from an enzyme called *aliinase*. Enzyme reactions are particularly fast. When the onion is cut, the aliinase swiftly breaks the chemical bonds of the sulfur molecule and forms new onion-flavored sulfur compounds that head straight for your tear ducts. So we have aliinase to thank for the onion flavor beloved of cooks and diners alike, but it is also the culprit that brings us to tears.

Knowing about aliinase is little comfort when your eyes are burning, and Newton's Apple wanted to find a way of keeping those nasty fumes from irritating our eyes. So we asked some volunteers to demonstrate their favorite fume-free techniques.

Ira: You're leaning over the onion with a

Though this volunteer said her eyes felt fine, Ira doubted it had anything to do with the clothespin on her nose.

burned match in your mouth. What's the theory behind the burned match?
Volunteer: The theory is that the charcoal is a good gas absorber and so it will soak up the onion fumes.
Ira: If you had enough activated charcoal the fumes would be absorbed, but one match doesn't seem to do the trick.
Volunteer: No, not very well.
Ira: Just judging from your very, very red eyes I'd say it's not a terribly good method. Let's try method number two. What is your theory here?
Volunteer: The flame from this candle causes an updraft that takes the onion fumes away from my face.
Ira: Is it working?
Volunteer: It's terrible.
Ira: There isn't enough of a draft to move the air. How about the next method—the old clothespin trick. How is it going? Getting any fumes in your eyes?
Volunteer: No. Last time I didn't.
Ira: I think that is just luck. It's your eyes, not your nose, you want to protect. So what is our next method all about?

"I'm inviting myself into the onion"

Volunteer: This is the Eastern method. I have a very sharp knife and I'm cutting along the natural lines or meridians of the onion in a gentle rocking motion. I'm inviting myself into the onion.
Ira: What a Zen-y idea! Be nice to your onion and your onion will be nice to you. But this onion is not welcoming you by any stretch of the imagination. I'd say this is a very angry onion. So I'll hasten on to our final method: the onion-in-a-tub-of-water technique. What's up?

Sure-fire cure: Eat it or cook it

Volunteer: I'm trying to keep the onion submerged in water so the fumes will dissolve in the water, not the air. So far so good.
Ira: The fumes are soluble in water, which means they are dispersed. This method works well. Here's our winner!

Two other methods are sure-fire: Eat it or cook it.

GOLF BY PROXY

Now pay attention and keep your eye on the ball. Iron Bobby, the new pro in town, is willing to spend all day helping his pupils conquer the greens. The important thing is good form, and while he may look a little stiff, Bobby's form is just great, right down to his toes. Remember, practice makes perfect—and best of all, no nasty cracks about your swing from this instructor!

Behind the Scenes 6

NEWTON'S APPLE

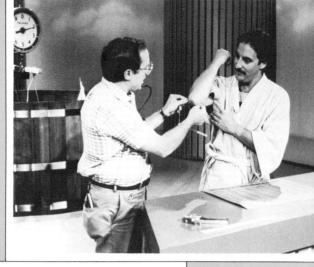

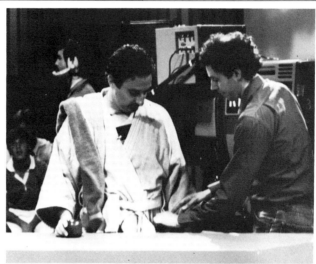

■ Sometimes when we start to plan a segment, there are technical problems to be solved that make a simple task formidable. This was the case when we decided to use a hot tub so that we could measure Ira's body fat. ■ The television cameras are attached to their electrical power source by long cables that snake over the floors as the cameras move about. Overhead there are banks of bright lights suspended from the ceiling. With all this electricity in use, we had to be very careful about introducing a large tub of water into the scene. ■ Fortunately all our precautions worked and we still have Ira with us. Unimpressed with this, he is still grumbling about the water being too cold!

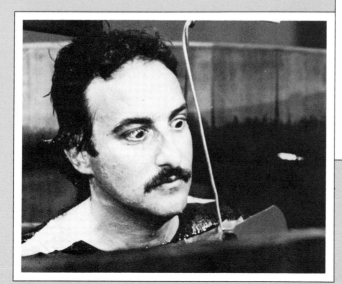

DEPTH IN SIGHT

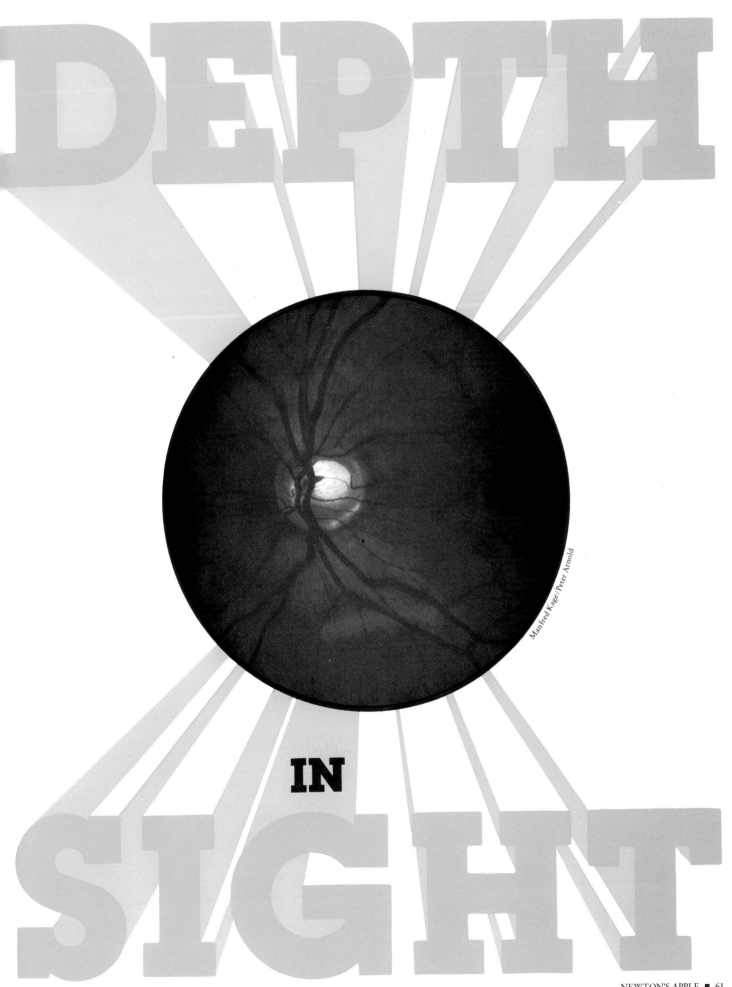

Manfred Kage/Peter Arnold

The compound eyes of insects, such as those of the fruit fly shown below, can provide stereo vision. But because these eyes are fixed in the insect's head, unlike the mobile human eyes shown on the opposite page, insects are unable to perceive dimension as humans do.

For two centuries some of the greatest minds have tried to understand how we perceive depth and why we are afraid of heights. Especially in the last two decades, researchers of infant development have tried to find out what newborn babies know when they emerge into the world and how they organize and use that knowledge.

Seeing the size and location of things in three dimensions is one of the most important adjustments we make to our physical environment. Many explanations are offered and experiments conducted, but we still have an incomplete understanding of depth perception and how it develops. How, exactly, do the mind and eye cooperate to perceive a three-dimensional world? Are we born with depth perception or do we learn it? When and how does fear of heights begin? These and similar questions about the way we learn to see the world are difficult to answer—but we're getting there.

The amazing thing about depth perception is that the eyes' receptor cells—called rods and cones because that's how they look in a microscope—are embedded in a curved, two-dimensional surface, the sideways- to front-looking eyes enabled us to see things against camouflage, a skill important to our survival.

Newborns have only rudimentary vision and would be classified as legally blind

🍎

Full depth perception seems to be the result of a long learning process. Babies emerge from the relative darkness of the womb with a rudimentary sense of vision, and they would be classified as legally blind. At eight weeks, the baby can differentiate between shapes of objects as well as colors (generally preferring red, then blue); at three months, it begins to develop stereoscopic vision. Yet even a two-year-old child who has learned much about the nature of his or her environment may still improve in depth perception.

We use many monocular (one-eyed) cues in perceiving depth:
• Linear perspective—as in the converging railway track illusion
• Atmospheric perspective—nearer objects are clearer than those far away
• Relative position—if two objects are in the same line of vision, the nearer one partly covers the farther one
• Light and shadow, or *chiaroscuro*—one pattern of light signifies "depression" and the other means "mound"

Humans are lucky their eyes face forward rather than sideways

🍎

retina. The word *retina* is derived from a word meaning "net" (of blood vessels). The retinas of our two eyes can work in tandem with the brain to synthesize their two somewhat different images into a single perception of three-dimensional space.

Humans are lucky that their eyes face forward and share the same field of view. Most vertebrate animals' eyes are positioned more toward the sides of the head, so they are aimed in opposite directions. As humans evolved, the gradual change from

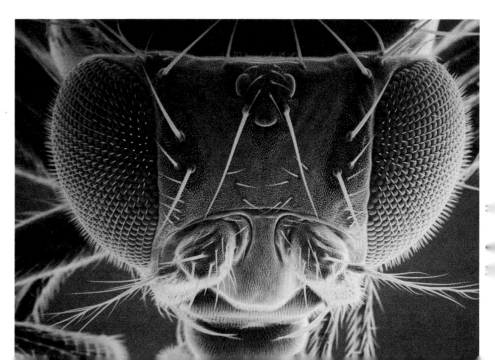

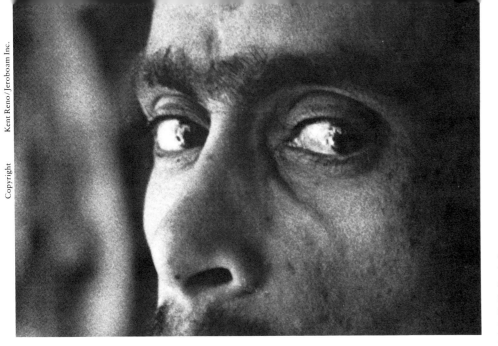

- Movement—near objects appear to move more rapidly than those farther away
- Accommodation—eye muscles contract or relax to accommodate the visualization of close or far objects

Retinal disparity, a binocular or two-eyed cue, is the most effective of all the cues to depth. The disparity of vision between the eyes separated from each other by a distance of about 2.5 inches creates slightly different images. You can easily test the disparity of perception between your eyes by looking at a fixed object with one eye and then, without moving your head, looking at the object with the other eye. The difference in the two images is greatest when the selected object is close, and the difference decreases as the object is farther away. The importance of the cue becomes evident when you try to thread a needle with only one eye open. Some art lovers look at paintings with only one eye. By doing this they eliminate retinal disparity and can appreciate the cues to three-dimensionality that the artist puts into the painting—assuming that the artist is trying to represent three-dimensional space!

Even though we experience depth mostly through our eyes, sometimes we obtain cues through temperature and other senses such as touch, smell, and especially hearing. When we listen to sound, the sound waves that reach our ears are slightly different in frequency, amplitude, phase and time of arrival. These slight differences provide cues to the sound source, which can be an effective aid to depth perception.

Eyes 2½ inches apart see slightly different images

BABY VISION

Since babies can't read letters on a wall chart, doctors are developing infant vision tests, using a flashing light. The baby's brain waves change in response to the light, indicating normal or abnormal vision.

A classic experiment for studying depth perception in humans and animals is called the "visual cliff." The apparatus looks much like a big playpen with an elevated clear plastic floor. On one side of the playpen a piece of opaque material under the plastic makes it appear to give solid support. The other side is transparent, so the infant or baby animal looks down at the floor about three feet below. When we placed two babies on the visual cliff, both of them crawled away from the deep side. In a film of the experiment with rats, kittens and a goat, all of the animals avoided the cliff.

An ingenious experiment to test depth perception in infants was thought up in the early 1960s by a woman named Eleanor Gibson while she was picknicking on the rim of the Grand Canyon. She wondered whether a young baby would crawl off the cliff or perceive the danger and avoid falling over the edge. She and R.D. Walk created a visual cliff simulation, and we have invited Dr. Albert Yonas, a professor of child development and a student of Eleanor Gibson, to repeat this famous experiment on Newton's Apple.

Ira: Dr. Yonas, are we born with a fear of heights?

Dr. Yonas: Infants are very good at acting pained when they are hurt or uncomfortable, but they are not very good at communicating that they are afraid. The very word *infant* derives from the Latin *infans*, meaning incapable of speech. Even so, the visual cliff test is helpful in determining depth perception in humans and other animals. The apparatus consists of a clear sheet of plastic under which a bridge divides what appears to be normal, solid support on one side from the other, which seems to drop off several feet.

Ira: At what age can you tell whether a baby has depth perception?

Dr. Yonas: One of the limitations of the visual cliff is that the baby has to be old enough to crawl.

Ira: Is there any way you can tell at an earlier age?

Dr. Yonas: Yes. At twenty weeks they reach for things that are closer rather than farther away. Even at four weeks of age an infant will blink as you bring something close to its face, indicating some depth perception. Children change dramatically over the first year, even though human babies develop much more slowly than other primates.

Ira: We decided to see how rats and kittens would act on a visual cliff, so we saw a film of rats, bred in total darkness and kept there for 27 days. When they were placed on the apparatus it was their first sight. All of them avoided crawling over the steep side. The

The goat wanted no part of the steep side

same thing happened with kittens reared in the dark. And the young goat we have on the show. . . .forget it. The goat wanted no part of the steep side. When such an animal is brought up on mountain sides, I guess it develops depth perception in a hurry.

Dr. Yonas: Some humans have a stronger disposition to fear heights than others. We don't know why this is, but it may be due to conditioning.

Ira: Seems like a good thing to have—that fear.

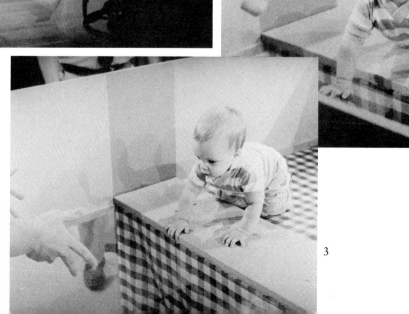

Wolves

Wolves are among the most maligned, yet fascinating, animals on earth. From a long line of horror movies they emerged to roam the landscapes of our imagination as sleek beasts lusting after blood. Superstition, folklore, legend and ignorance have contributed to the wolf's low reputation, which is almost totally undeserved. It is because of the fear and revulsion wolves inspire, rather than their occasional attacks upon livestock (or their even more rare attacks on people), that these gray hunters have been exterminated from most of their range.

Wolves once inhabited all of Eurasia, from India to the Arctic snowfields. Although still common in many parts of central and northern Asia, the wolf has disappeared from most of Europe. The beleaguered creature is extinct in eleven European countries: Ireland, Great Britain, France, Belgium, The Netherlands, Denmark, Switzerland, Austria, Hungary, and both of the German states. And they are very scarce in North America and Canada.

Small wolf populations survive in the Iberian peninsula, Italy, Bulgaria, Czechoslovakia, and Poland, as well as in the European part of the U.S.S.R. The wolf remains rather common in the tundra of

The wolf's low reputation is undeserved

🍎

northwestern Russia, the Urals, the mountains of the Balkans, and especially the Caucasus, which has a wolf population about as dense as any other part of the globe. More than 4000 wolves roam the Caucasus Mountains. In the southern hemisphere the maned wolf, whose long legs give it great swiftness, lives in the Pampas regions of Argentina and Uruguay.

Wolves in the Carpathians, and in the Balkans as well, tend to migrate up and down the mountainsides according to the season. During the summer, when herdsmen drive cattle, sheep and goats into the higher mountains, the wolves follow. This is the season when wolves are most inclined to prey on livestock. When the snow piles up on the peaks again, the domestic herds are brought down from the mountains. The snow also drives the wolves to lower heights, where they hunt their natural prey, deer and wild boar.

We asked Nancy Gibson, our fountain of knowledge from the Minnesota Zoo, to bring two wolves here and talk with us about them.

Ira: Nancy, why do wolves have such a terrible reputation?

Nancy: I think that Little Red Riding Hood got to us early in our lives! More seriously, wolves compete for some of the food that humans eat, so wolves have the reputation of being vicious.

Ira: Well how do wolves fit into the wild? What is their niche in the ecology?

Nancy: They are at the top of the food chain. They'll eat large hoofed animals like moose and deer. They have also been known to take down bison.

Ira: How do they manage to take on as large an animal as a bison?

Nancy: They can do it because they hunt in packs. Perhaps that's another reason that humans have this eerie response to them. They are social like we are. Anyway, communication between social animals is very important to them. Their howling, which contributes to their eeriness, is part of their communication.

Ira: Why do they do that?

Wolves used to roam over much of our globe, but human predators have drastically reduced their numbers. Now they are an extremely endangered species, even though in the wild they generally go out of their way to avoid humans.

Howling is a social activity that builds rapport and can be heard for many miles

Tom Cajacob/Minnesota Zoological Garden

Nancy: It can mean several things. It builds a rapport within the group. Whenever they start to howl, they gather together in a pack. Sometimes they do it before they go out for a hunt. It's a communications system that can be heard for miles and miles.

Ira: What kind of social hierarchy do wolves have, if any?

Nancy: They have quite a strict hierarchy, headed by an *alpha* (meaning "dominant") male and female responsible for breeding and gathering together of the pack. The alphas initiate the hunt, and are often first to eat the prey. The average pack is made up of about eight members, and has a social structure in which every individual finds its own level. The order of precedence is accepted by the community, but sometimes they fight for the leadership.

All wolves have excellent equipment for hunting. First of all, they have large ears for hearing prey moving through the woods. And they have very large feet that function like snowshoes for traveling long distances

---

Large feet function like snowshoes

---

in search of food. They also have long legs for traveling through snow, because most of them live in northern regions where snow is abundant. Their large feet function just like snowshoes. Other aspects of their hunting gear include good eyesight, a good sense of smell, and massive skulls with strong muscles to power the large canines with which wolves tear open the flesh of their prey.

Ira: Do they make good pets?

Nancy: No. They are animals meant to exist in the wild. They are an endangered species and should be left alone to propagate in the wild. Besides, would you want an 85-pound beast roaming around your house?

Ira: You've got me there, Nancy.

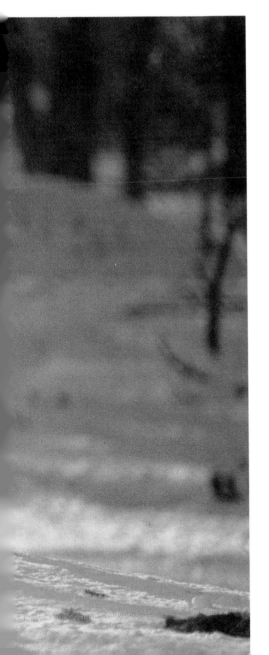

Why Do Our Ears POP?

Copyright   Kent Reno/Jeroboam Inc.

Ira: Imagine yourself in an airplane that is taking off or landing. Your ears pop, and it can be very painful. You can do several things to try to avoid it. And here to tell us about popping ears are Jan Serie and her eerie friend, Dead Earnest.

Jan: Dead Earnest is here to show us his Eustachian tube, which connects his throat to his ears. This tube is really the key to explaining why your ears pop. Two of the three chambers of the ear are connected with ear popping. The outer one, your ear canal, is the one you're not supposed to put anything in.

Ira: I was told you should never put anything smaller than your elbow in your ear.

Jan: Good advice. Anyway, the other chamber is called the middle ear, and it is separated from the ear canal by the eardrum, or the tympanic membrane, which is the technical name for it. Your middle ear is attached to your throat by the Eustachian tube.

Ira: What happens when you go up or down in an airplane and your ears pop or hurt?

Jan: Your Eustachian tube normally is closed, but when you go up in an airplane you bring along sea-level pressure in your middle ear.

Ira: It stays trapped in there?

Jan: That's right; it's trapped. In the outer ear the air gets thinner and thinner and so you have less pressure. So the denser pressure in the middle ear bows the eardrum out into the outer ear.

Ira: Ouch! So you feel pain in your eardrum because the high pressure pushes it out?

Jan: Yes. And to alleviate the discomfort you can balance the pressure on both sides by swallowing. A little muscle attached to the Eustachian tube lifts the tube and allows air from your mouth to flow into or out of your middle ear. Then the pressure on both sides is the same.

Ira: So you equalize the pressure by swallowing or chewing gum to lift the Eustachian tube? Then the pressure goes down and the pain goes away?

Jan: The pain goes away, and that's when you hear the pop.

Ira: Ah! The pop!

Jan: Yes. It pops because the bowed-out eardrum snaps back into place. When your eardrum moves, the movement is sensed as a pop.

Ira: When you have a cold, it's much harder to pop your ears. Is that because there's something wrong with the membrane?

Jan: No, it's just inflamed and clogged with mucous. You can open the Eustachian tube but air can't get in or out. That's why it really hurts when you have a cold and fly in an airplane.

Ira: Thanks, Jan, for explaining how airplane flights can make you play pop goes the eardrum.

POCKET SKI LIFT

An inventive Austrian has come up with a pocket ski lift. Roll the rear wheels over the transmission device and then hike up your favorite slope to set up the anchor post, pulley, and length of steel wire. No waiting, no being restricted to a tow-equipped hill. Your jalopy works while you play. Although it's made of wire, you do need a special grip to hold on to the cable. Spinning wheels sing a song of freedom and frolic for the owners of the pocket ski lift. It's a little ski whiz!

National Archives

THE

How do doctors find out how much stress the heart can take? Newton's Apple asked an expert on cardiology, Dr. Jay Cohn, to explain how a stress machine works.

Ira: Tell me, Dr. Cohn, what does this machine do?

Dr. Cohn: It does a lot, Ira. It not only helps to diagnose the presence of heart disease, but it monitors the response to therapy, it makes decisions about surgery, and it prescribes exercise programs for people who

Finding out whether your heart function is limited

are being rehabilitated from heart attacks and other cardiac problems.

Ira: You have me hooked up to this machine for a good reason, I hope.

HEART HOOK-UP

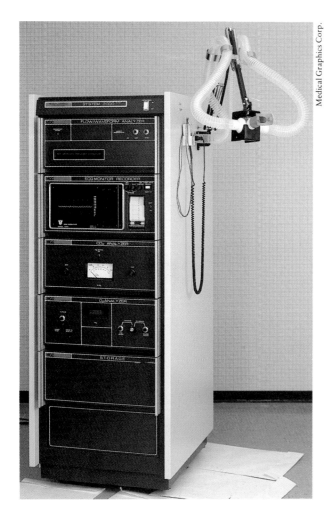

Medical Graphics Corp.

As Ira peddles, the heart machine (shown at left) measures how much stress his heart can take while exercising. The machine monitors blood pressure, pulse rate, amount of air breathed, and amount of carbon dioxide and oxygen exhaled. From this information doctors can learn how efficiently a heart delivers blood to the muscles, which directly affects the muscles' ability to do work. An individualized exercise regime can be accurately prescribed using the data from this machine.

Dr. Cohn: We've got you hooked up to this gadgetry because what we're going to do is to monitor how your heart functions during exercise. You know, when you're at rest your heart is working at less than 20 percent of its capacity. So when we exercise you, we force the heart to do more. Thus we can find out if your heart is functioning normally or whether your heart is limited in what it can do. And this gizmo can help to find out. Our technician, Marcus Mianulli, will hook you up to the machine.
Marcus: OK, now, Ira. You shouldn't breathe through your nose or talk. And good luck to you.
Ira: OK, I'm ready to roll.
Dr. Cohn: Now start exercising while I point out all the things we're measuring. First of all, we have a blood pressure cuff

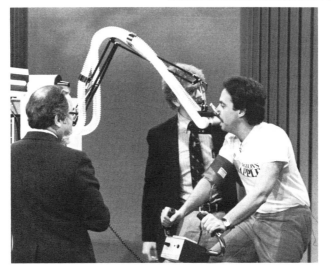

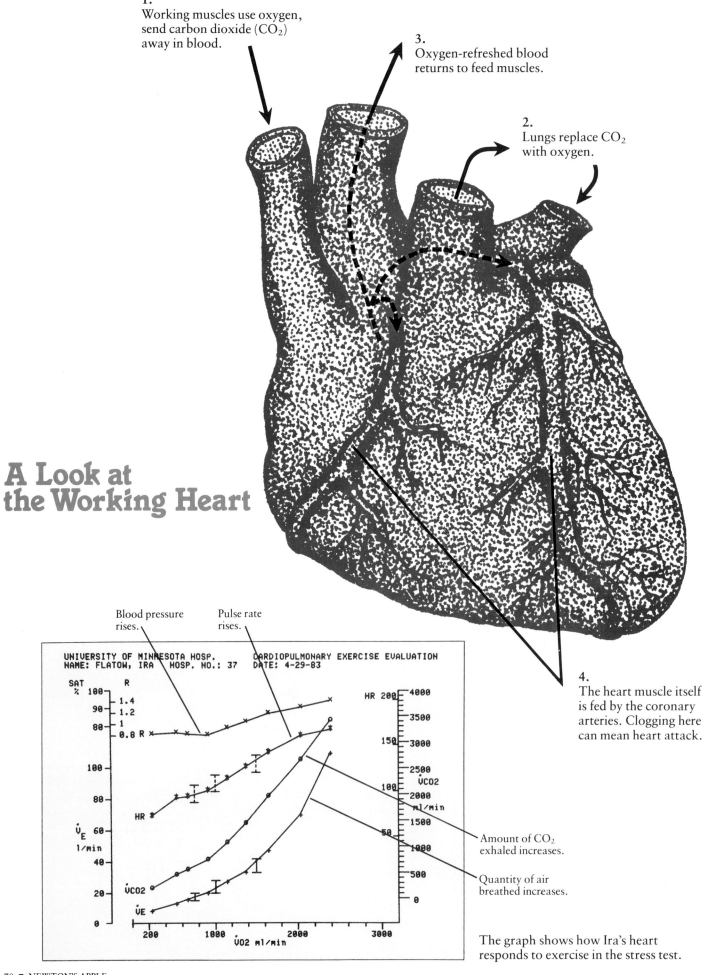

A Look at the Working Heart

1. Working muscles use oxygen, send carbon dioxide (CO_2) away in blood.

2. Lungs replace CO_2 with oxygen.

3. Oxygen-refreshed blood returns to feed muscles.

4. The heart muscle itself is fed by the coronary arteries. Clogging here can mean heart attack.

Blood pressure rises.

Pulse rate rises.

Amount of CO_2 exhaled increases.

Quantity of air breathed increases.

The graph shows how Ira's heart responds to exercise in the stress test.

around your arm and we're recording your changes in blood pressure. Next, we are recording the amount of air you're moving in and out of your lungs with every breath. We're also taking an electrocardiogram of your heart. Over here we are recording the carbon dioxide and oxygen concentration of the air that you expire with each breath. By monitoring this concentration we can determine how much oxygen your muscles are consuming as you exercise. And that oxygen has to be delivered from the heart. So this measurement indicates how efficiently the heart is delivering blood to the muscles and providing a sufficient amount of oxygen so the muscles can do work.
Ira: (Huff. Puff.)
Dr. Cohn: Well, Ira, let's unstrap you and let you talk again.
Ira: Boy, it's not an easy thing to take this test. It really does stress your body. So tell me what all these little lines mean?
Dr. Cohn: The different colored lines are variants of the measurements of your oxygen consumption. They show the oxygen consumption at the point where you couldn't go any longer. That was a very good exercise test that you did.
Ira: What does it tell you about me?
Dr. Cohn: You consumed enough oxygen, certainly, to be normal. But you're not well-conditioned.
Ira: Oh boy, the secret's out. I'm out of shape!

At rest, your heart is working at less than 20 percent of capacity

Tom Murray

HEARTBEATS

When you hear someone's heartbeat, you're not actually hearing the heart muscle beating. What you hear are the valves in the heart slapping shut.

Behind the Scenes 7
NEWTON'S APPLE

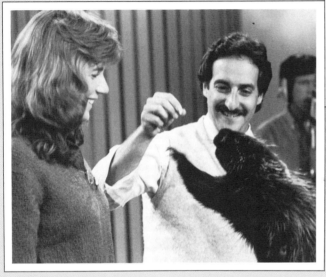

■ Our studio crew especially likes the animal segments because they have an unusual opportunity to be close to the wild animals. Everyone was a little nervous about touching or holding the porcupine, though, because of its sharp quills and claws. Suddenly the porcupine astonished one of our associate producers by jumping onto her back! Ira came to the rescue, lifting the friendly little creature away. ■ All during its visit, the porcupine kept chewing away at the corners of the table. After we finished taping the segment, we repaired and repainted the table. Perhaps we should have left the evidence. Very few people can say that a porcupine has nibbled on their table.

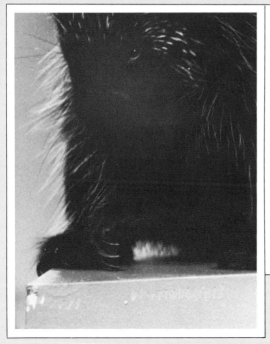

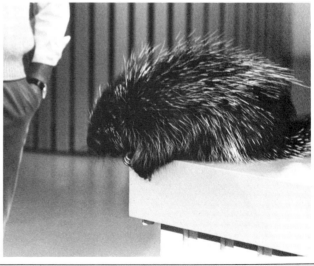

A WORLD OF

COLOR NOT SEEN

For most of us, the world abounds in glorious color, but some people lack the ability to see a full range of colors. This deficiency can have far more serious consequences than the inability to choose the appropriate necktie for a certain outfit. Color pervades most aspects of our lives, and what we accept as a given is actually one of our most useful tools for survival. Color provides essential cues to health, food quality, weather and many other elements of concern as we maintain ourselves in balance with our environment.

If your eyes are color deficient, some objects become indistinguishable, unidentifiable or invisible, so that adapting to your environment can be difficult and frustrating. Color deficiency might be compensated for more readily were it not for the fact that color is frequently employed to make warnings more visible and to distinguish among options in all sorts of human activities. Lack of adequate color perception can interfere with the educational process, prevent people from entering certain careers, cause the loss of a job or a transfer, contribute to accidents, and—perhaps saddest of all—limit enjoyment of Nature's beauty.

Color deficiencies are passed on from one generation to the next through the chromosomes. *Chromosomes* are bodies within cells that carry *genes*, infinitesimal bits of material that determine the traits you will inherit from your parents. Each person has 22 pairs of chromosomes, plus a pair of sex chromosomes. Males have 44 chromosomes plus an XY pair of chromosomes; fe-

More serious consequences than choosing the wrong tie

GENES and VISION

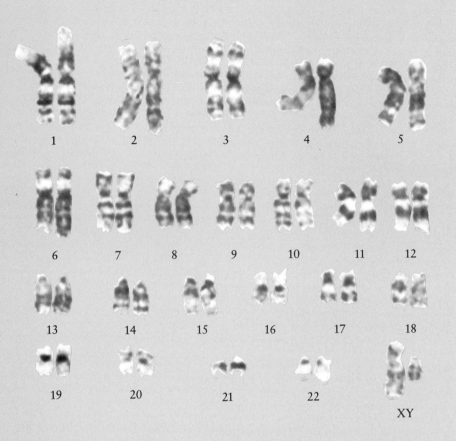

National Institutes of Health

Pages 73–76 compare the way objects appear to people with normal color vision and those with the two most common deficiencies: the inability to perceive red and the inability to perceive green. When the red light cannot be seen, the flowers appear to be dark yellow or black. When green is unseen, colors appear to be saturated or washed out. Color deficiencies are inherited through genetic material composed of chromosomes (above) carrying genes composed of DNA (below).

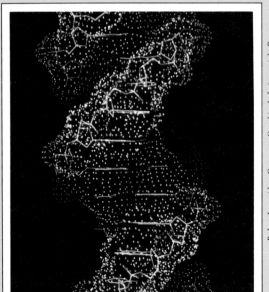

Robert Langridge, Computer Graphics Lab./copyright Regents, U. of California, S.F.

Fundamental Photos

males also have 44 chromosomes, but with an XX pair of sex chromosomes. The male's Y chromosome is inert, so female cells have 46 functional chromosomes and male cells have only 45. This difference of one X chromosome containing hundreds of genes determines male and female characteristics.

Males exhibit the trait and females transmit it

The absence of a functional mate to the X chromosome in males has genetic consequences, allowing recessive genes or sex-linked genes (as in color deficiencies and hemophilia) to exert their effect. In females, these undesirable effects may be masked by a dominant gene of their other X chromosome.

The most common sex-linked human trait is red-green color deficiency. It is mainly inherited by sons through their mothers; males exhibit the trait and females transmit it. Because males possess only a single X chromosome, the incidence of color vision deficiencies is greater among males than among females (8 percent, or about 8,500,000 males, as opposed to 0.5 percent, or about 400,000 females). Unfortunately, scientists have not devised a test to determine a carrier of color vision deficiencies.

Trichromatic, or normal, color vision is an uninterrupted perception of the visible

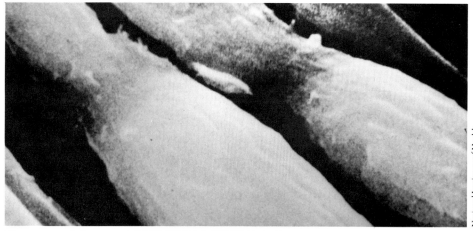

The cones of the eye.

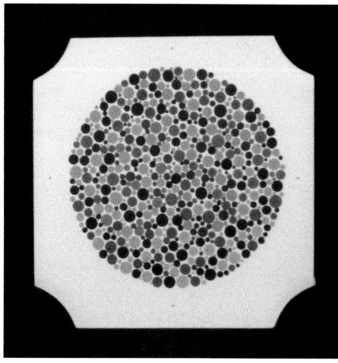

Can you see the numerals?

Color information is processed by three cone systems, or *photoreceptors*, located in the back of the eye.

light spectrum, without black, gray or white areas. Those with normal color vision can differentiate among the six or seven hues of red, orange, yellow, green, blue and violet. Any of these colors can be produced by mixing the three primary colors of red, green and blue, which is why normal color vision is called *tri*chromacy (three colors). Trichromats can differentiate hundreds of colors of varying hue, brightness and saturation.

Abnormalities of color vision are classified as *anomalous trichromat* and *dichromat*. The former is less severe and more common. Anomalous trichromats can match colors with red, green and blue, but require the presence of more than the usual amount of one of the occurring colors. People with this affliction have a weakness in one of their color systems, so they need different proportions of the primary colors than are required by those with normal color vision.

Dichromats are color-defective individuals who require only two primary color stimuli to match all other colors. The three types of dichromats are *protanopes* (lacking red), *deuteranopes* (lacking green), and *tritanopes* (lacking blue). Red and green types of deficiencies are the most common.

Color information is processed by three cone systems, or photoreceptors, located in the back of the eye. Three visual color pigments in the cones of the retina are sensitive to red, green and blue. Each cone has a separate sensor, so different cones are stimulated by different wavelengths of light (that is, different colors). The cones react to the primary colors and send messages to the brain that mix them in appropriate proportions to provide normal color vision. In a person with color vision deficiency, the color pigments of the cones are changed, and hence the perception of color changes. The mechanism for the defectiveness of the pigments in the cones is not completely understood.

Color vision deficiency can be recognized by its symptoms or by one of several color-deficiency tests. It is not as apparent to the person having it as one might think. John Dalton (1766-1844) was an English chemist who first described color blindness in a publication in 1794, but he did not recog-

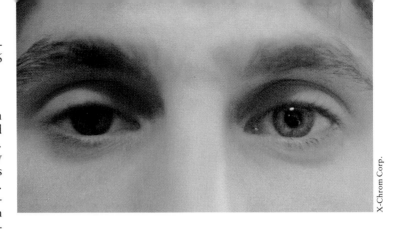

publication in 1794, but he did not recognize his own color defect until he was 26 years old.

At present no cures exist for color vision defects, but one recently developed aid, called *X-Chrom*, appears to be helpful. X-Chrom is a lens that was developed by Dr. Harry Zeltzer as he was studying means of improving underwater vision for divers. Applying his findings to color vision deficiencies, Zeltzer discovered that a lens with a ruby-red hue was effective for the common sorts of red and green deficiency. The lens increases the number of shades that a color-deficient person can see. Once these new shades have been properly identified, color-deficient people can learn to recognize colors they never knew existed.

One eye is always dominant in humans, a situation analogous to right- or left-handedness. Zeltzer found that the dominant eye, the one without the X-Chrom lens, receives partially incorrect color messages. The X-Chrom contact lens on the dominant eye sends improved messages to the brain—not unlike seeing the world through one rose-colored glass. The lens does not change or eliminate the vision defect, but it can help individuals to adjust better to a full-color world.

For the color deficient, one ruby-red lens

SEEING RED

Contrary to popular belief, bulls are not attracted to the matador's red cape because of its color. Bulls cannot see the color red. They respond to the *waving* of the cape regardless of color.

The Tongue and Taste

UPI

A well-known Latin epigram says *de gustibus non disputandum est*, or "there is no point in arguing about taste." It is not as well known that the sense of taste probably differs quite a lot from one individual to another. The intricate biochemical processes of tasting and smelling, combined with the complex chemistry of the things we taste, cause countless variations in the way we perceive them.

How many categories of tastes are there? Through the ages many people have ventured an opinion.

"Two," said Aristotle, identifying them as sweet and sour.

"Seven," wrote Father Polycarpe Poncelet in a book on the chemistry of tastes and odors published in 1755. He listed acid, sweet, bitter, peppery, stale, astringent, and sweet-sour.

"Four," say most modern authorities: sweet, salty, bitter, sour.

"Six," maintain some experts, who believe that alkaline and metallic tastes can be detected too.

Only when foods are mixed with liquids can they be tasted. A solid in a dry mouth creates no sensation of taste, for taste buds are stimulated only when chemicals in food are dissolved by saliva and washed over the taste buds. The tongue contains about 10,000 nerve endings, our taste buds, which are located in minute protrusions called *papillae* (from the Latin for

---

A solid in a dry mouth is no taste sensation

NEWTON'S APPLE ■ 77

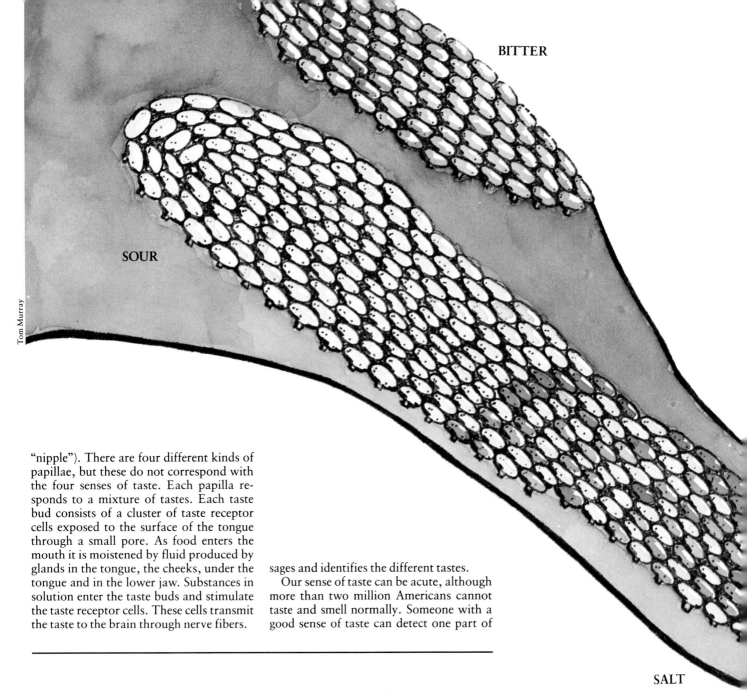

"nipple"). There are four different kinds of papillae, but these do not correspond with the four senses of taste. Each papilla responds to a mixture of tastes. Each taste bud consists of a cluster of taste receptor cells exposed to the surface of the tongue through a small pore. As food enters the mouth it is moistened by fluid produced by glands in the tongue, the cheeks, under the tongue and in the lower jaw. Substances in solution enter the taste buds and stimulate the taste receptor cells. These cells transmit the taste to the brain through nerve fibers. sages and identifies the different tastes.

Our sense of taste can be acute, although more than two million Americans cannot taste and smell normally. Someone with a good sense of taste can detect one part of

Taste buds grouped on the tongue: bitter at the back and sour on the sides, but salty and sweet at the tip

Most of our taste receptors are concentrated on the upper surface of the tongue, but a few are located on the soft palate and on the epiglottis, the thin piece of cartilage in the throat that prevents food from being inhaled into the lungs. The four different types of taste buds lie on different parts of the tongue. They are not uniformly distributed over the surface of the tongue; instead, they are grouped. Salty and sweet sensations are tasted best at the tip, bitter ones at the back, and sour ones at the sides. There is little sensation in the middle of the tongue. The taste buds transmit messages to the brain and the brain sorts out the messages a chemical among many thousands of chemicals. But when we get older, many taste buds disappear, along with the appreciation of flavors. That may explain why some parents may disbelieve their child's complaints about foul-tasting food or medicine.

The tongue also contains nerves that register temperature, pain and touch. All of these influence the way things taste. For example, most people agree that a hot cup of tea or coffee tastes better than a cold one. Consistency, texture and appearance also influence our attitudes toward food and its appeal to us.

The senses of taste and smell constitute a unity such that, when we eat, neither sense can function independently. People who cannot smell can taste only sweet, sour, salty and bitter, but not any of the subtle flavors like those of chocolate, roast beef, lemon or herbs. An overwhelming majority of people who complain that they have lost their sense of taste usually have

Sweet tooth: Vestigial means of survival?

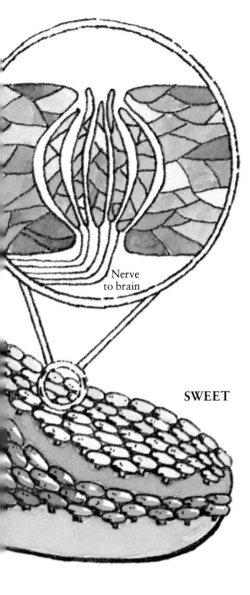

SWEET

tooth; we certainly show a preference for sweet foods early in life. If babies two or three days old are offered liquids of varying sweetness, they suck hardest and longest on the sweet-tasting solutions. That initial preference for sweets can be reinforced by the culture. Some countries have a tendency to consume more sweets than others. The average Briton eats more sweets than anyone else in the world.

One explanation for the love of sweets is that the body needs sugar in order to make glucose. But this may be only part of the answer. In the jungle, bitter-tasting plants and berries are frequently poisonous, whereas sweeter ones are usually safe to eat. So a sweet tooth may be a vestigial means of survival.

something wrong with their ability to smell, not taste. Food flavors, which are primarily smells rather than tastes, are perceived differently. When food is put in your mouth, odors travel up the back of your nose until they reach the olfactory (smell) receptors. So if you want to savor something in your mouth you should inhale, which forces air from the mouth across the olfactory receptors. If your olfactory receptors are damaged, say, by a blow to your nose, they can repair themselves. They are one type of nerve that regenerates itself once every month or two.

It seems that we may be born with a sweet

Why Does Your Hand FALL ASLEEP?

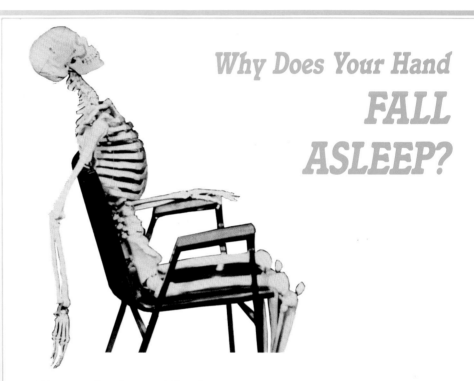

Ira: When your hand or arm falls asleep, it really can be painful. But why is that? Here to answer that question is Jan Serie. Jan, it's good to see you and Dead Earnest again. He's your favorite companion, isn't he?
Jan: Not really. My other friends are more talkative. But I brought him with me to demonstrate what happens when your hand falls asleep. As you can see, we've dressed him up a bit with some blood vessels and nerves.
Ira: He looks 100 percent better.
Jan: The blood vessels and nerves go all the way down his arm to his hand. Your hand falls asleep when you put pressure on this blood vessel. The pressure cuts off the blood supply to the nerves. And if they can't get any oxygen from the blood, they won't work. They won't send any sensation back to your brain, and they won't allow your muscles to move. That's when you feel the weird sensation we commonly call "asleep."
Ira: Sometimes that happens to me and I get scared. I might wake up in the middle of the night and become afraid that my normal sensations are never going to come back. Numbness forever.
Jan: Usually it hurts when this happens, and the pain wakes you up. But a condition called Saturday Night palsy can be serious.
Ira: Is that a new dance step?
Jan: Say you go out drinking and get a bit inebriated. If you sit in a chair whose back is under your armpit, and if you fall asleep in that position, normally it starts to hurt and you wake up. But if you're too inebriated to wake up, then this can go on all night. Eventually you kill the nerves in your arm so they will not work at all.
Ira: You have permanently paralyzed your arm?
Jan: It's not permanent. The nerves slowly grow back. But it can take about six months before you feel sensations in your arm again and can move it correctly.
Ira: This is not something that we sober folks need to worry about.
Jan: No.
Ira: Be careful to keep Dead Earnest off the sauce. We need him for the show.

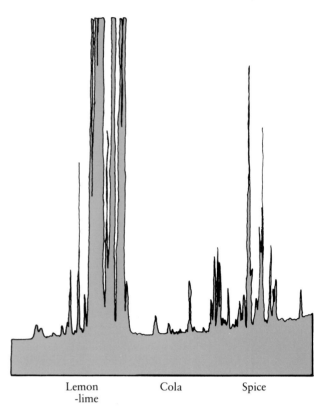

A taste of Pepsi

Newton's Apple wanted to know why people tasting the same food can respond in totally different ways. So we invited some tasters from the audience to compare the tastes of Pepsi and Coke. We also invited Dr. Jan Serie, a biologist with impeccable taste, to teach us about the tongue.

Ira: Welcome, Jan. While our taste panelists are tasting away, can you tell me why I hate canned spinach and other people like it?

Jan: Did you have a bad childhood experience with spinach, Ira?

Ira: I sure did. I was forced to eat the stuff.

Jan: Well, it also had something to do with your taste buds. And a lot of other factors influence how you taste something. To show you what I mean, take a taste of this ice cream cone.

Ira: Yech! That's no ice cream cone. It's mashed potatoes.

Jan: Don't you like mashed potatoes either?

Ira: I love mashed potatoes. When I first tasted this, though, it didn't taste like mashed potatoes or ice cream. I couldn't tell what it was.

Jan: You expected ice cream, but the texture and smell were wrong, so even the taste on your tongue was wrong. Your expectation got in the way of what you were really tasting.

Ira: So a lot of taste is in your head?

Jan: Sure. Take coffee. A lot of people don't like coffee when they start drinking it. But if you think you are going to like what you are about to taste, you are much more likely to like it.

Ira: That's all news to me. Let's go over to the panelists and see how well their taste buds are working. Panelist 1, are you finding a difference between Pepsi and Coke?

Panelist 1: I found that the Pepsi has a fuller flavor, a sweeter taste, and more carbonation. The Coke seemed more watery and has an acid aftertaste.

Ira: Panelist 2, what's your opinion?

Soft drinks contain only half the chemicals in a cup of coffee

Panelist 2: I disagree with Panelist 1. The Coke is sweeter and has a caramel flavor. It has some kind of exotic flavor that I can't quite discern. The Pepsi is more effervescent and has a citrus flavor.

Ira: It's interesting that you two had opposite tastes. As for you, Panelist 3, you look very familiar.

Panelist 3: Yes, Ira, I'm the ringer on the panel. I'm a trained taste tester.

Ira: Aha! It's no surprise that taste is very important to food companies. I suppose that texture and appearance are important to the business, and I know that food companies have separate research departments to determine which foods will be gobbled up by consumers.

Panelist 3: You bet. And the difficulty in taste testing a sample of people is that it is difficult to translate what you sense into words.

Ira: Well, how would you differentiate the Pepsi from the Coke?

Panelist 3: I agree with some of the comments the other panelists have made. Pepsi has a more distinct flavor. It is lighter and has more carbonation. Also, it has a lemon-lime character to it.

Ira: Thank you, panelists, for your "Taste

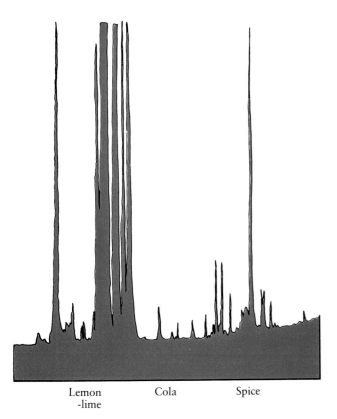

Lemon-lime   Cola   Spice

A taste of Coke

Ira, Dr. Gary Reineccius and Jan Serie discuss the workings of the gas chromatograph. The readouts to the left show comparisons of the flavors that make up the tastes of Coke and Pepsi.

and Tell." And now we're going to get a little more scientific about the difference between Pepsi and Coke. We have with us Dr. Gary Reineccius, a flavor chemist. Dr. Reineccius, tell us about this machine.

Dr. Reineccius: This is a gas chromatograph, which is an extremely sensitive apparatus for analyzing the components of a complex mixture of substances. To show how it works we put a little Coke in the machine so that it can smell the aromas. Once all of the liquid is separated into its chemical components, the gas chromatograph prints out a graph, called a *chromatogram*, of all that it has detected.

Ira: And these are all the different chemicals that are sensed by the machine? Seems like quite a bit.

Dr. Reineccius: Yes, somewhere between 600 and 700 chemicals, which is only about half of what you get in a cup of coffee, or a third of what is in bread. If you look at the chromatogram you can see that Coke has a large peak, which indicates the presence of ethanol, a form of alcohol. It is used in trace amounts in processing the lemon and lime. There are lemon and lime in both Pepsi and Coke. And 7-Up too.

Ira: Jan, are you impressed with what this machine can do?

Jan: I'm definitely impressed, but it can't tell us whether it enjoyed drinking the stuff it has examined. It takes the human being a lot of faculties—taste, smell, prior experience, anticipation, appearance and so on—to put together the perception of pleasure or displeasure with food. The machine has its limitations.

Ira: Dr. Reineccius, you're not going to stand there and hear your machine maligned like this, are you?

Dr. Reineccius: No, I'm not. Suppose we want to monitor the flavor quality of milk in a dairy. This requires tasting it every 10 minutes, 24 hours a day, 7 days a week. A person could get very tired of doing that. Of course, the machine can't replace the abilities of the human being, but it can be useful in certain quality control situations such as this.

TERMITE TRICKS

Termites may have invented chemical warfare. When a termite bites an enemy, it applies a toxic anti-healing chemical to the wound. Termites also brush contact poison onto the surface of their attackers with their enlarged upper lips and spray their predators with an irritating entangling agent.

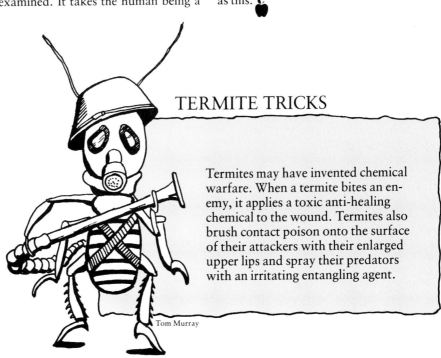

Tom Murray

BLACK HOLES IN SPACE

"Good news today," Albert Einstein wrote to his mother in September 1919 upon hearing the report of the eclipse expedition. For two years, while war ravaged Europe, the Royal Astronomical Society of England had forged ahead with plans for its critical experiment to test Einstein's General Theory of Relativity. In this theory, Einstein showed that the gravitational field of massive bodies actually warped the surrounding space. Motions of objects in the warped space follow curved paths; in everyday terms we would describe these motions as "falling" due to gravity. But the crucial new element in General Relativity was that anything that moved must follow these curved paths, even light. This predicted behavior reinforced the notion from Einstein's Special Theory of Relativity that mass and energy were equivalent, a concept embodied in the famous equation $E = mc^2$.

Einstein predicted that if the light from a star passed near the edge of the sun on its way to us, the warp in the starlight's path would show up as a slight shift in the star's apparent position. But the only time one could observe the stars near the sun was during a total solar eclipse. Thus, expeditions were sent to Africa and Brazil to observe the eclipse of May 29, 1919. The investigators compared their eclipse photographs with photographs of the same star fields, taken six months earlier. They found position shifts of approximately 1/3000 degree (the size of a dime seen from a distance of 1½ miles). This result confirmed Einstein's predictions, and revolutionized the world of physics.

One consequence of the falling of light due to gravity was the possibility that a sufficiently massive body could actually trap light. On the Earth, for example, objects are trapped if their speed is less than the so-called escape velocity of approximately 7 miles per second. If the escape velocity from some massive object were greater than the speed of light, 186,000 miles per second, then nothing could escape, not even light. The result would be a black hole.

The concept of black holes is not new; it was proposed almost a half-century ago, but most physicists ignored the proposal until the 1960s. By then, theoretical physicists, using General Relativity, showed that black holes were not just a theoretical possibility, but might actually form from the sudden collapse of stars.

To understand this, consider a large and massive star. The size of the star is dictated by two opposing forces: the outward force produced by nuclear fusion at the core that tends to blow the star apart, and the inward force of gravity that tends to pull the material of the star together. The two forms balance one another. But when nuclear fuel in the star is spent and fusion diminishes, the tug-of-war is no longer balanced—gravitation predominates and the star undergoes gravitational collapse.

Matter does not fall directly into the black hole. Instead, the black hole may revolve, whereupon a ring of gases forms around the hole. This ring, or *accretion disk*, constantly feeds the hole while emitting x rays and gamma rays. The detection of this radiation leads to locating and identifying a black hole.

We wanted to find out more about these mysterious black holes, so Newton's Apple invited Dr. Lawrence Rudnick, an astronomer, to visit with us.

Ira: Welcome, Larry. Let's get right to the point. What is a black hole?
Larry: A black hole is an object in outer space that is so compressed that its gravity does not allow anything to escape from it—no particles and no light.
Ira: It's hard to understand what you mean when you say that not even light can escape from something.
Larry: Light is bent by gravity, as this apparatus can illustrate. This lens shows what would happen if the moon turned into a black hole. If it were a black hole, it would be just as big as this little spot.
Ira: That little spot on the side of the lens?
Larry: Yes. When light tries to go around this spot, the light is bent closer to the black

Gravity becomes so strong not even light can escape

hole. It gets bent more and more until, finally, it is bent so much that it never escapes from the black hole.
Ira: How does a black hole form?
Larry: We believe that black holes can form naturally as a star goes through its life. I'll give you a simple demonstration. A star is maintained by inside pressure that resists its gravity. Gravity is trying to pull the star inward, but the internal pressure from its molten gases offsets the gravitational pull. This balloon will give us an idea how that works.
Ira: This is our star?
Larry: Let's imagine that it is. The surface tension in the balloon is trying to pull it in, but the internal air pressure holds it out and, when it is heated, even causes it to expand. But if I take that heat away by pouring liquid nitrogen over the balloon, look what happens.
Ira: It collapses! That's amazing. And the same mass of air is still inside?
Larry: The mass is there, but much condensed. Just like the balloon, the star would collapse as it cools.

Ira: Of course we can't see a black hole, right?
Larry: Not directly.
Ira: But you could detect it by other events surrounding it?
Larry: Yes. Primarily because it distorts the space around it.

A glowing effect modifies our image of these holes as entirely black

The image of these holes as entirely black was revised about ten years ago by Stephen Hawking, a brilliant British physicist who is one of the prime theorists of black holes. He proved mathematically that atomic particles inside a black hole find their way out of the black holes and evaporate. This activity creates a kind of glowing effect. That is, a black hole, when running backward in time, can create a white hole, according to Hawking's proposition. He went on to suggest that the universe is full of mini black holes, which means that matter other than massive stars can create black holes. This suggestion affects the way we describe how the universe was created and what its future will be. In effect, his proposal may mean that there is an ebb and flow in the energy balance of the universe.

Whatever the physical facts of black holes may really be, we still do not know whether the universe will end with a bang or a whimper—or, indeed, if it will ever end at all.

Behind the Scenes 8

NEWTON'S APPLE

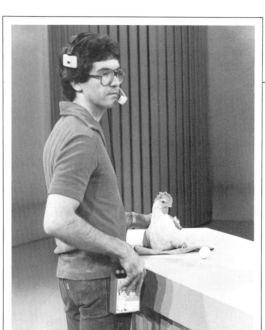

■ We now know that one of the hardest things to do on television is to try to direct a chicken! When we did the segment on eggs, the chicken was totally uncontrollable. Desperate for a solution to our problem, we took seriously a suggestion to try hypnotizing it by swinging a pencil in front of its eyes. To no one's surprise, it didn't work. ■ Finally we placed the segment in the hands of fate and left the chicken in the studio overnight, hoping it might warm up to its surroundings. As far as we can tell, the chicken still doesn't like TV.

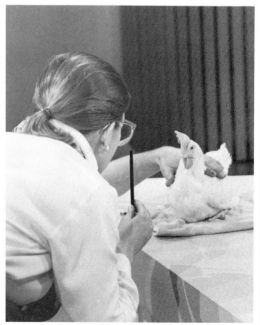

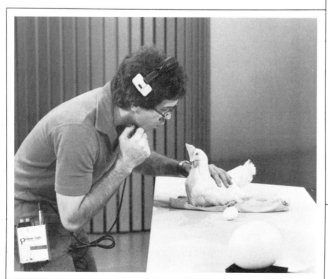

What is Digital Sound?

Our musical instruments have invariably reflected the most advanced technology of the time. So it is not surprising that the phenomenal advances in integrated-circuit technology have been incorporated into modern musical instruments. The revolutionizing aspect of the digital music synthesizer and the digital recording of music is enormous. It is having a profound effect, comparable to the introduction of the violin or the piano in historical times. But this is just the beginning. Eventually, radio and television will be transmitted digitally to provide high-quality, interference-free reception.

We wanted to learn more about digital sound, so Newton's Apple invited David Stringer of TV Ontario to tell us about it.

Ira: Dave, what is all this noise about digital sound?
Dave: Do you remember painting by numbers? Digital sound is sound by numbers.
Ira: Sound by numbers. Sounds like an interesting concept, but what does it mean?
Dave: We need to compare sound in a digital system to sound in an analog system. The analog system has its name because in such a system analogies are constantly being made. Sound in an analog system is best represented as squiggles showing wave forms of the sound, whereas in a digital system a list of binary numbers (ones and zeros) are all we need to faithfully reproduce the sound that the numbers represent. In sampling a digital sound, computers recognize about 50,000 numbers each second. Even though there are still gaps in this kind of recording, they are so tiny that they are invisible, if one can call sound invisible.
Ira: Why would you want to change an analog signal into a group of numbers?
Dave: There are definite advantages to doing so. Do you know the party game called "Telephone," where you whisper a message down a row of people and by the time the message reaches the end it's usually very different from the original? Well, analog equipment tends to add a little bit of its own personality in the form of hisses, static and other noise. If you played the same party game with computer equipment by writing

The medium no longer affects the information

the numbers down on a piece of paper, the paper might be smudged at the end but you would still be able to recognize what the numbers are, and it doesn't matter if any static has crept into the message.

Digital sound, then, is sound converted to computer numbers, and it is the most dramatic change in sound production in one hundred years. So let's look a little more closely into the evolution of digital sound.

In 1928, when a scientist named Harry Nyquist began the digital revolution, 78-rpm records were still being improved. But Nyquist proposed a theory, called the Signal Sampling Theory, that would be an alternative to the traditional method of transmitting and preserving sound. Instead of copying waves in wax, vinyl, magnetic particles, or electrical current (the analog approach), waves could be sampled at regular intervals and their voltages noted. These samples might then be read back to replot the original wave. The information would be more accurate in this form because the medium in which it is presented no longer affects the information itself. For example, the stylus of a phonograph cannot distinguish between a recorded wave and dust on the playing surface.

Nyquist went on the suggest that at least twice the total of the highest frequency recorded was necessary for faithfully reproducing sound. Because human hearing can generally perceive about 15,000 hertz, acceptable results could be achieved with upwards of 30,000 samples per second.

In 1928 no system existed that could measure, record, store or reread 30,000 samples in a second. However, a partial solution came in 1937 when G.R. Stibitz built the first digital relay computer from telephone switching equipment. It was from Stibitz's idea to use binary arithmetic (the indications of on and off by ones and zeros) to store and manipulate great volumes of information that digital got its measuring scale and its name.

Another twenty years passed before digital computers possessed the speed and capacity to respond to zeros and ones in the needed quantities. But the invention of the transistor in 1947 made large-scale digital transmission a likelihood. For the first time a plausible means was created for all those zeros and ones to be collected and read. They were strung together as multidigit "words" to represent each microsecond's sample. By the early 1950s, analog-digital and digital-analog converters had been produced that could generate 30,000 of those words each second and convert them back into sound. In the late fifties the invention of the integrated circuit permitted real-time digital transmission.

Continued on page 88

Eastman Kodak

Above is a highly magnified photograph of the grooves on a traditional phonograph record. It is an example of an analog system of recording. At the top of the opposite page, a digital musical score is shown. Below it is a photograph taken by a scanning electron microscope that shows how the binary number system is simplified on digital recording tape to a series of on and off signals. The disk in the person's hand is a digital recording disk, which is impervious to the wear and tear that ruin the sound quality of analog recordings. At the bottom right is Herb Pilhofer composing with digital equipment. Below can be seen a cumbersome analog computer from days gone by.

analog

Bell Laboratories

86 ■ NEWTON'S APPLE

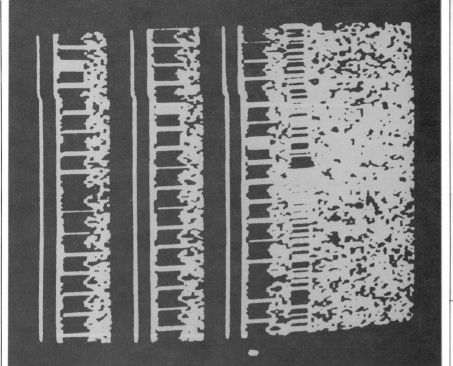

Digital

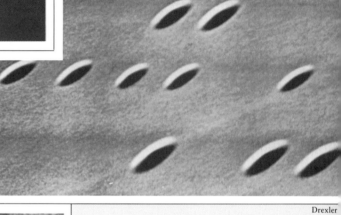

Drexler

Newton's Apple visited the composer Herb Pilhofer, who showed us what he can do with a digital synthesizer.

Herb: Unlike analog systems, the digital synthesizer can change speed and pitch independently, because once a series of notes is put into the computer it becomes numbers, not sounds. And so I can manipulate it any way I want. For instance, I can take any part of a melody and loop it. In other words, I can take just part of the digital information and turn it into a rhythmic pattern.

Ira: Not only can you manipulate the sounds by ear, but the computer provides you with a graphic representation of the note. The computer can present the notes in several forms—even a printed score.

Herb: What I like about this whole setup is that it frees me from the process of writing on a piece of paper, having someone copy it, and then having somebody else play it two weeks later. I can now create things—assign a sound, a color, to it, and instantaneously react to it—all in one sitting. I can change, manipulate, or do whatever I want. Now I can be a painter of sounds.

Continued from page 85

Ira and David Stringer discussing digital sound equipment.

Computer-generated music still has numerous obstacles to overcome. Because computers can produce *any* sound, it is often difficult to produce a musically useful sound. Many composers think of music as organized sound, and they want to find new organizations for new sounds. This can be a great challenge for the listener to hear it as order rather than chaos.

Since 1960 Max Mathews, affiliated with Bell Laboratories' acoustical research center, has been experimenting with computers and digital-analog converters to create the sounds of a new music. His work has inspired many avant garde composers and scholarly institutions. He believed that there is a great potential for nonprofessionals to make music at home. Accordingly, he came up with a program called CONDUCTOR that should thrill any music buff. A score can be put into the computer and the user can conduct it: rebalance the instruments, change timbres and dynamics, adjust tempos—just as if standing in front of a symphony orchestra.

Electronic instruments have continued to enjoy great popularity. Many traditionally acoustic instruments have an all-electronic version: the piano, organ, accordian and guitar, to name a few. In fact, the electronic versions of the organ and guitar are commercially far more popular than their acoustic antecedents. It is difficult to make a precise distinction between these electronic instruments and a music synthesizer.

A Computer program lets you be the conductor

The situation is analogous to the distinction between calculators and personal computers. Just as personal computers are more versatile than the calculator, the electronic music synthesizer is more versatile than electronic instruments.

Computer-produced sounds and electronic sounds in general have strongly influenced nonelectronic music, though computer music has not yet become popular on a large scale. However, its influence is particularly evident in relation to the ever-more-demanding requirements of sound quality. Unlike the unpleasant sounds of early electronic music, contemporary computer music integrates sounds from a

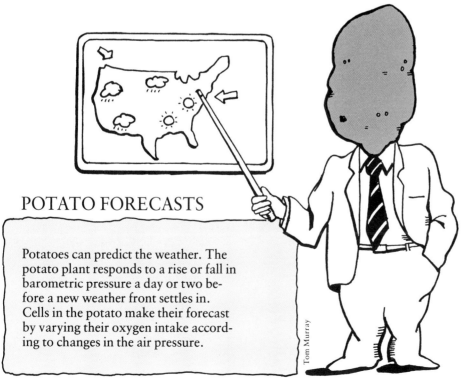

POTATO FORECASTS

Potatoes can predict the weather. The potato plant responds to a rise or fall in barometric pressure a day or two before a new weather front settles in. Cells in the potato make their forecast by varying their oxygen intake according to changes in the air pressure.

MACHINE FOR WALKING ON WATER

This man thinks he can walk on water. And he can! The original model was engineered in Bavaria. It may draw a few smiles, but it works! Up, down, up, down. All it takes are high aspirations and a little muscle power.

variety of sources, and the result is usually more pleasing.

Choosing 50,000 sample amplitudes a second, we can produce a sequence of 50,000 short electrical pulses a second in

"Now I can be a painter of sounds"

such a way that the amplitude of each is equal to the sample amplitude. This train of pulses is passed through a filter that eliminates any frequencies above 25,000 hertz. Because we are free to choose any numbers that give the sample amplitudes in any way desired, we can produce any possible sound wave whose bandwidth is 25,000 hertz or less. This is good enough for high-quality musical sound.

Sounds can be processed and recorded on a digital audio disk that plays one hour on each side. The information is stored as zeros and ones, but they can't be seen because they are so tiny. These numbers are read by a laser beam that reflects light and plays the corresponding sounds. A digital audio disk is virtually indestructible because the information is recorded beneath the surface of the disk. So your digital recording of Bach may be nearly as timeless as the composer himself.

NEWTON'S APPLE ■ 89

Bactrian Camels

Ira: Look what we have here! A baby camel from the Minnesota Zoo, brought by Nancy Gibson. Welcome, Nancy and Kevin Dale, the camelkeeper. I have to say, this doesn't look like a camel I'm used to seeing. I mean, I don't see a hump.
Nancy: Well, they are here, underneath this hair. Two humps! Most people associate camels with the great big one-humped variety, called dromedary camel, which is an Arabian camel from the desert regions. This two-humped camel is called a *Bactrian* camel because it comes from Bactria in the Mongolian Steppe regions.
Ira: Why does it have all this hair on it?
Nancy: So it can live in cold climates as well as warm. The seasons change drastically in Mongolia.
Ira: This one is just a baby.
Nancy: Three months old. Her humps will grow as she grows.
Ira: How big is she going to get?
Nancy: She's going to get pretty big. She will weigh about 1500 pounds, which is the average weight for an adult female camel. The males can reach 2000 pounds.
Ira: Let's talk about some camel myths. First of all, the humps. Are they used to store water, like everybody thinks?
Nancy: No, that's a common misconception. They store fat in them. If necessary, the fat in the humps can be metabolized into water. But they generally carry the fat for insulation and as an energy reserve.
Ira: We hear that camels can go for months without food. Is that because of the humps?
Nancy: They don't go without food for months, but they can go for a few days. During the breeding season the males lower their food consumption and use their humps for a source of food. When the breeding season is over, their humps sometimes lie almost flat. But the most amazing thing about camels is their adaptability.
Ira: Adapted for the way they live?
Nancy: Bactrian camels are from Central Asia, an area with extreme hot and cold weather, and they've got several wonderful features for survival in such harsh climates. First of all, they have a double set of eyelashes that keep out wind, snow and sand. Also, their nostrils are closeable, to keep out the elements. And they have cloven hooves that allow them to walk over rocky terrain and sand or snow. Camels amble, which consists of moving same-side legs forward while the legs on the opposite side are moving backward. Then they have this incredible hair! In the summer they shed it and are quite bald. Then in the winter they grow back a thick winter coat to keep them warm.
Ira: Do camels live in other parts of the world?
Nancy: Well, there's the South American camel, the llama.
Ira: The llama's a camel?
Nancy: The llama is in the camel family along with the vicuna, alpaca, and guanaco.
Ira: What does this camel eat in the wild?
Nancy: In the wild she'll eat a grain mixture, some grasses. At the zoo we feed this baby supplemental milk. I'll give you the honor of feeding her.
Ira: She's drinking from this bottle very, very thirstily. We're going to run out of milk in a second. Is this natural milk that she's drinking?
Nancy: No, this is just evaporated milk. How are you at baby burping?
Ira: I think I'll pass on that! 🍎

Not water, but fat in those humps

Double sets of eyelashes keep out wind, snow and sand

HOLD THE VEGGIES

Cats fed a purely vegetarian diet will go blind. To maintain their eyesight, cats need taurine, an amino acid found in meat. People, dogs, and other animals can synthesize taurine from vegetable proteins. Cats can't.

COOKING TH

It has been said that one true test of a good cook is a perfectly boiled egg. It seems a simple thing to do, yet all kinds of complications can develop. Soft-boiled becomes hard-boiled the minute you turn your back. You want to take deviled eggs to a picnic and the yolks turn out green. To produce a dozen uncracked Easter eggs for decorating, you go through two dozen, or more. Ignorance of what happens during the boiling process has led many an egg boiler to simmering frustration. What does go on inside this secret world when it boils?

Maybe the act of boiling an egg strikes you as an uncomplicated event, but the transformation of an egg's biochemical system is an intricate one, as you would expect when you consider that the system was intended to nourish and create a chick.

Probably the most noticeable change is that the boiling egg's contents are changing from a semifluid state to a firmer, but still resilient, state. As this happens, subtle changes take place at the molecular level. For one thing, the white part, or *albumin* (from the Latin word for "white"), breaks weak chemical bonds and forms stronger ones. The albumin is composed of about one-tenth protein molecules and nine-tenths water. Ordinarily coiled up, the protein in the albumin opens up and begins to form a three-dimensional protein network as the egg heats. This network is more rigid

---

Albumin breaks weak chemical bonds and forms stronger ones as it cooks

🍎

and less mobile than the protein molecules were in the semifluid state. As the temperature rises, the network expands, which can lead to one of egg-boiling's biggest heartbreaks: cracking.

What causes an egg to crack? There are several reasons and several remedies. Every egg has an air sac at its larger end. When the pressure builds up inside the egg, air is released from the sac through the small pores at the larger end. But if the air sac heats up faster than the air can leave the shell, the internal air pressure will crack the egg. The best technique for controlling air flow is to stick a pin in the egg's larger end. This pinhole allows the air to escape and thus produces a perfectly formed egg without a flat bottom.

Eggs will crack if they are placed directly into boiling water. When they are started in cool water, their internal temperature increases gradually, so that the egg's air has more time to seep out and the albumin doesn't become firm before the yolk is cooked.

A major source of grief in egg-boiling is the mess that happens when the white leaks out through the cracks. A homespun remedy for this is to add salt or vinegar to the water. Salt cannot get into the eggshell. Instead, it helps to seal off cracks from the outside, because the electrical charge of the

E GOOD EGG

salt molecules can quickly break apart the protein's bonds and link the protein to the more rigid three-dimensional protein network. So the protein does cook faster and better if salt or vinegar are present to seal the cracks.

Another egg-boiling disaster that fries most people is when the egg ends up with an ugly green yolk. As the egg boils, some of the protein's sulfur and hydrogen atoms form hydrogen sulfide gas, the infamous smell of rotten eggs. This hydrogen sulfide gas moves to the cooler region of the egg, which is the yolk. Because yolks contain iron atoms, which have a strong affinity for bonding to sulfur atoms, the hydrogen sulfide gas quickly converts to gray-green iron sulfide. The remedy is to plunge the boiled egg into cold water as soon as you remove it from the heat. This will keep the hydrogen sulfide gas away from the yolk and result in beautiful yellow yolks.

Many an egg boiler simmers in frustration

Here are some practical tips for boiling eggs:

- Begin cooking eggs in cold water.

- Use fresh eggs because their air sacs are smaller than those of older eggs; scramble the older ones.

- Boil hard-cooked eggs for 10 to 12 minutes in water that is at a gently rolling boil

- After boiling, immediately run cold water over the eggs.

Why Do Teeth CHATTER?

Ira: We want to know why your teeth chatter when you're cold, so we asked Jan Serie to drop by with Dead Earnest, who looks like he could use some dental work.
Jan: As you know, Ira, when you are cold you shiver.
Ira: Do I ever. Florida, here I come.
Jan: The muscles that move your jaw also shiver when you are cold. When they shiver, the two muscles that move the jaw, the masseters, start to twitch and move your jaw in little spasms. As the masseters twitch, your teeth move up and down rapidly, making a chattering sound.
Ira: Why does the body shiver?
Jan: Shivering is an effective mechanism for raising your body heat. If your body loses heat, it has to take measures to keep you warm. One good way of keeping warm is to exercise. When the muscles contract, it warms you up. Shivering makes use of muscle contraction without causing you to move. It's mostly an involuntary movement.
Ira: Some people shiver because they're scared, not cold. Is the same mechanism at work?
Jan: No. That kind of shivering is not understood very well. Some think that when you're scared you prepare yourself to fight or flee, depending on the situation. Your body is telling your muscles to get ready to do one or the other of those two things. But if you're not ready to do either, your muscles will just twitch and shake.
Ira: Thanks for telling us, Jan. Maybe you should get Dead Earnest a coat or something. It might warm him up—and would certainly make him less scary looking.

Behind the Scenes 9

NEWTON'S APPLE

■ When the robot named HERO 1 was being delivered to the studio, the strangest thing happened. As it was being lifted out of the back seat of a compact car that had transported it to our studios, it suddenly said "Please don't bother me. I am trying to sleep."

THE ROBOTS ARE COMING!

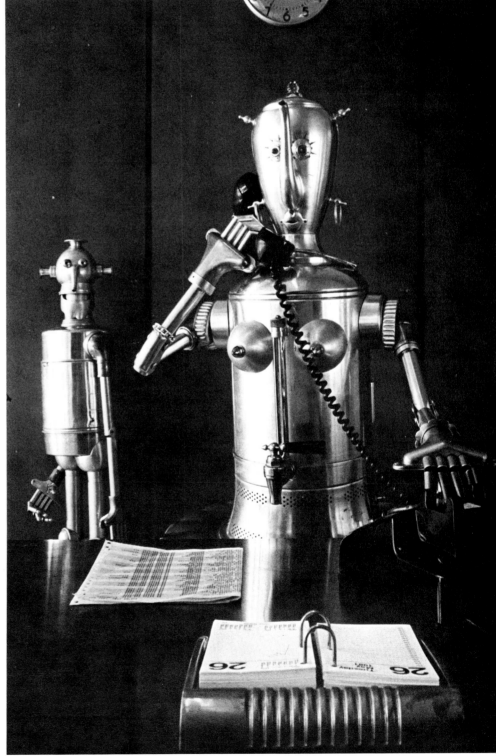

Long before the book of Genesis told of God's creating Adam and Eve, myths about the making of human beings abounded. We have never lost our preoccupation with the synthetic creation of human life. One of the most popular expressions of that preoccupation is *Frankenstein* (1818), Mary Shelley's story about Victor Frankenstein's attempt to build a living creature. His motives for making what turned out to be an abhorrent monster echo the desires of others who have wished to recreate life: "I thought that if I could bestow animation upon lifeless matter, I might in process of time renew life where death had apparently devoted the body to corruption."

A popular preoccupation with the synthetic creation of life

The word *robot* was not invented until 1921, when playwright Karel Capek wrote *R.U.R. (Rossum's Universal Robots), A Fantastic Melodrama*. Since then, the word has been variously defined and interpreted. But Isaac Asimov's classic book *I, Robot* (1950) stated three immutable laws of robotics:
1. A robot may not injure a human being, or, through inaction, allow a human being to come to harm. 2. A robot must obey the orders given it by human beings except where such orders would conflict with the First Law. 3. A robot must protect its own existence as long as such protection does not conflict with the First or Second Law.

Now that we have entered the age of computers, the making of robots is a reality. We are on the brink of a technological revolution (which some call *mechatronics*) that represents the next quantum jump in the evolution of machines. What makes this revolution particularly interesting is that amateur experimenters, rather than industrial or university researchers, are taking the initiative in building better robots. Robotics is a convergence of several fields,

Robot: Clayton Bailey/Photo: Don Peterson

all of which revolve around automation and computer systems. The field is open to all comers.

Even though Asimov's three laws of robotics are helpful in defining what a robot is, or should be, there are lingering questions about what constitutes a *true* robot. Most robotics experts agree that a robot must be an autonomous machine; that is, it must be capable of carrying out functions on its own, which differentiates it from a typical computer system. Given a command by a human operator, a true robot must be able to execute the command by freely deciding how to do it.

Robots show great promise for improving industrial efficiency and quality performance, but they have raised issues as to their effect on human beings in the workplace. Will robots put large numbers of people out of work? Will workers managing robot systems perhaps find their jobs more challenging and interesting? Because both the technology and utilization of robots are still in their infancy, such questions are difficult to answer. It is clear, though, that robots can relieve workers of difficult and dangerous physical work.

At this stage in their development, robots are limited in their abilities. As Hans Moravec, a leading expert in robotics, remarked, "Computers are at their worst trying to do the things most natural to humans, like seeing, hearing, language and common-sense reasoning. Large portions of our nervous systems are dedicated to these skills, and computers are unlikely to match human performance until they can process as much data as the neural centers. I calculate that a typical present-day computer is a million times less powerful than a human brain. The computer can perform a million single steps per second, but the 40

billion neurons in the brain can switch a thousand times in the same interval. The spectacular, continuing evolution of microelectronics should make the requisite power available at great cost in a decade, and inexpensively by the end of the century."

In short, robots will need to have much more highly developed abilities to sense, to move, and to think before they can take over more complex tasks in factories. At present they are able to perform only simple tasks such as loading and unloading parts,

Personal robots equipped to make small talk

drilling holes, inserting screws, painting, welding and other comparatively limited functions.

Industrial robots aside, there is now a boom in personal robots comparable to the boom in personal computers. Personal robots usually have on-board computers and primitive sensors that allow limited movement around the home—as well as some small talk! Only a handful of personal robots, or *homebots*, are on the market, but they are finding many eager buyers. Most personal robots today are little more than fancy toys that can perform only the simple tasks they have been programmed to do. Even so, they can be a useful tool in learning about computers.

Newton's Apple wanted to get better acquainted with the family of personal robots, as well as some industrial ones, so we invited Dr. David Thornburg, an authority on robotics, to introduce us.
Ira: Welcome, Dr. Thornburg. You must feel quite at home amidst all these robots.
Dr. Thornburg: This is robot heaven for me.
Ira: These robots seem to come in all shapes and sizes. But what about this one here? It doesn't look like a robot at all. More like a toy.
Dr. Thornburg: It is a toy. This is the Milton Bradley "Big Track." But it is also a robot because it is a machine that moves according to controls from a computer that is built into its body. I can show you how it works by writing a program for it, just as we would for a large robot.
Ira: A little computer is in there?
Dr. Thornburg: Yes, and the robot is remembering all of the instructions I type into the computer. He remembers all of the steps.

Ira: What an age we live in!

Dr. Thornburg: This other robot, named HERO 1, is built by the Heath Company. It's an extremely educational robot (HERO stands for "Heath Educational Robot") because it allows people to control not only its arm, but also its movement around the floor. It does all sorts of things, from sensing sounds to speaking, to locating and picking up small objects. But its primary purpose is to educate its owner in robotics.

Ira: HERO 1 seems to be a step up from Big Track.

Ira: They seem to be very accurate at what they do. I mean, they know exactly where to go to pick things up and put them down. How do they know that?

Dr. Thornburg: They know because a human being took the time to put these robots through their paces. As that is happening, the robot's computer keeps track of all of that information, and so every time you want to do something over again, you just tell the computer to repeat it. Of course if there wasn't anything there for them to work with, the robots would still be spray-

Not as smart as we may think

🍎

robots are the personal robots people can use in their own homes. Here is one of my friends. His name is Topo.

Ira: He actually looks like a robot.

THE ROBOTS ARE COMING!

Dr. Thornburg: It's quite a step up. The Big Track stores only sixteen instructions, whereas HERO 1 can store many more than that.

Ira: And he's got an arm and a wrist that move. Do you have to teach him what to do?

Dr. Thornburg: You teach him with this controller, and then he remembers all the instructions and carries them out.

Robots that educate their owners about robotics

🍎

Ira: That's amazing. These things are nice, and fun, but they're essentially toys, aren't they?

Dr. Thornburg: They're certainly not as practical as robots used in industry.

Ira: We have a couple of those right here. They seem to be big and noisy, though. What are they doing?

Dr. Thornburg: They're very noisy, but they can't hear, so it doesn't matter to them! This robot is carrying pieces from a work station to the other robot, which is meticulously applying glue to the pieces. This is just one of the many kinds of applications for industrial robots. They can be used for spray painting, welding, working in dangerous environments such as steel mills—all sorts of things.

Ira: They do take the place of people, then?

Dr. Thornburg: That's right. Who wants to do this sort of thing? There are enough dangers in life without getting yourself killed in the workplace.

ing glue and trying to move things.

Ira: So they are not as smart as we may think they are.

Dr. Thornburg: No, absolutely not.

Ira: Because they work so well in assembly lines and monotonous jobs, I imagine we will be seeing more and more of these robots used in industry.

Dr. Thornburg: That's right, Ira. But even more exciting to me than these industrial

ROBOTS

The word *robot* comes from the Czechoslovakian *robota*, which means "work" or "compulsory service."

Dr. Thornburg: He does. He looks just like what people think robots should look like. Topo was made in California by a company named Androbot, which was started by Nolan Bushnell, the inventor of Pong. Topo is radio-controlled by an external computer, and if I give him a command he'll carry it out.

Ira: Whoops! Watch out! I thought he was going right over.

Continued on page 100

NEWTON'S APPLE ■ 97

At the left, an industrial robot that weighs three tons can pick up and move delicate or fragile objects with great finesse. Below, a tabletop model is used to perform intricate tasks in circuit design. Below left, RB5X is a home robot that can bring you the morning paper. On the opposite page, rented robots gather at a convention, and one named Argon, a professional companion, goes for a walk with the neighborhood kids.

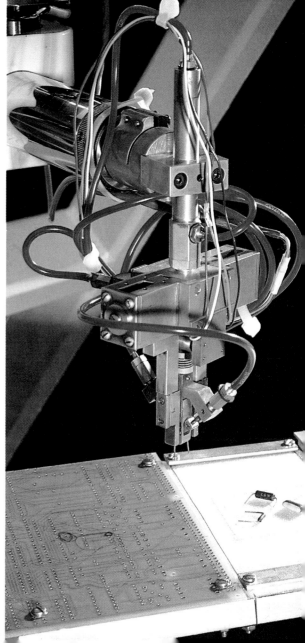

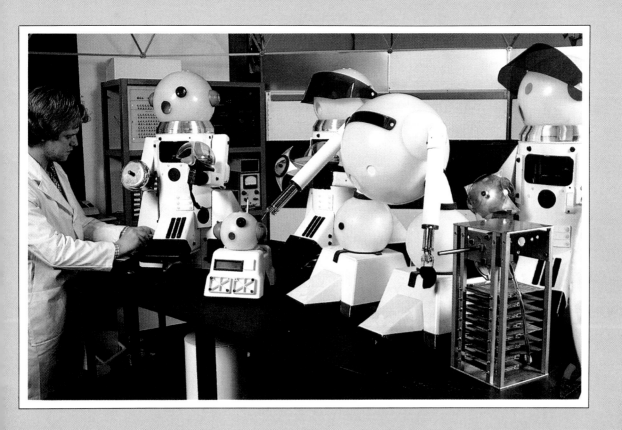

THE ROBOTS ARE COMING!

Continued from page 97

At right are HERO 1 and RB5X. HERO 1 is intended to educate its owner in robotics; its arm enables it to pick up and put down small objects. Armless RB5X stands in the background. Below, Dr. David Thornburg is introducing Ira to the other guests on the show. Bottom right, a desktop Teachmover, an inexpensive way to learn the rudiments of robot operation; this can be programmed simply by rehearsing its movements through space.

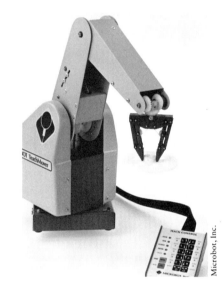

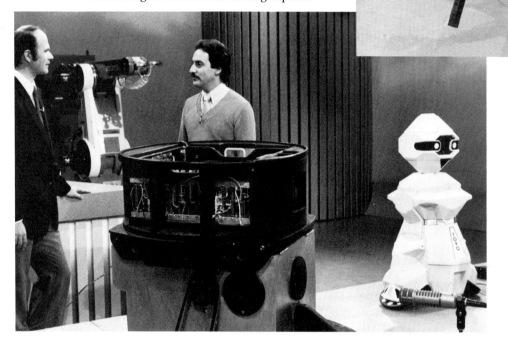

Dr. Thornburg: No, Topo would never go off the deep end, because I programmed him not to do that.
Ira: He didn't know enough on his own not to go over there?
Dr. Thornburg: No, not yet. That's coming, though, and it is important that we develop intelligence in computers if we want them to work in the home.
Ira: But if you want him to be more autonomous—to work on his own—he's got to be able to sense where he is.
Dr. Thornburg: Yes. He has to have sensors. These are very important because homes are designed for people, not robots. People can move in complex ways that robots have not been able to copy yet.
Ira: So they have to know where they are to get around the furniture and other obstacles.
Dr. Thornburg: Sure. The most important thing is for a robot to sense its surroundings. Here's a robot that does that. His name is RB5X and he is made by RB Robots in Golden, Colorado. When I get him going, you'll see that if he bumps into something, he senses that and turns around.
Ira: Does he sense things with those little white squares?
Dr. Thornburg: Yes. These are just little bumpers, little switches. They are quite sophisticated as far as switches go, but very crude when you think of all the other things they could be sensing. This has a sensor for measuring distance. It's similar to the sensors on instant cameras in that it uses sound waves.

Little white squares that send out sound waves

Ira: It sends out a sound wave that bounces off something and that is how it knows it's near something and changes direction.
Dr. Thornburg: You could also use infrared energy to sense the presence of human beings, and you could use speech recognition to understand peoples' voices.
Ira: The autonomous robot. That seems to be what you're really aiming for, isn't it?
Dr. Thornburg: I'm not sure. It's not at all clear to me that robots should function totally on their own.
Robot: We'll be right back, Ira.
Ira: I see what you mean.

Among this family of robots and their younger siblings, additional functions are being developed to make them a more helpful part of the household: vacuuming floors, washing windows and dishes, entertaining guests, detecting smoke and intruders, fetching things. Many experts predict that by the turn of the century homebots will be handling most household chores.

Will robots eventually take over the world, as some fear? It's difficult to say, but there is no doubt that they'll become smarter and smarter. Hans Moravec says of the future of robotics and artificial intelligence, "Machines as intelligent as humans will, in their generality, be capable of superhuman feats and will be able to do the science and engineering to build yet more powerful successors. . . . It is clear to me we are on the threshold of a change in the universe comparable to the transition from non-life to life."

Have you hugged your robot today?

What Fat?

Fat is a bad word to most Americans, and we spend lots of time and money trying to get rid of it. But a certain amount of fat is necessary to our health and a normal part of our body makeup. Even though we weigh ourselves on the bathroom scales and anxiously compare our weights to charts of the established norms, few of us have a clear idea of what fat is and how much of it is on our body.

Fats are chemically simple compounds made of carbon, hydrogen and oxygen. One molecule of fat has two components: glycerol (a sweet, sticky, odorless and colorless liquid that is used in the manufacture of products ranging from explosives to antifreeze) and one or more compounds called fatty acids. When the glycerol and fatty acids combine, a fat molecule forms.

People rarely stop to think of the positive aspects of body fat. Body fat cushions the organs and the joints that are subject to pressure. It also serves as insulation. Fats constitute your body's most highly concentrated source of energy, and are utilized for survival when food is not available. Fats provide far more calories (a measurement of heat energy) per weight than either proteins or carbohydrates, because they require less water for storage and conversion to energy.

Half the body's fat reserve can change location daily

Fat is tucked away in various places throughout your body. It can be found in the connective tissue beneath your skin, around your muscles, and surrounding your organs. About half of your stored fat is deposited just under the skin in what is called *subcutaneous tissue*. This layer of tissue attaches the skin to underlying tissues and organs. The fat is dispersed around the kidneys, the stomach and intestines, in genital areas, and between muscles. Fat is also stored behind the eyes (of all places), as well as in the furrows of the heart.

Surprisingly, fat does not remain in one spot until it is needed for energy. It is continually released from storage, transported through the bloodstream, and redeposited in other fat cells. As much as one-half of the total body fat reserve changes location daily!

We wanted to know how to determine how much of a person's weight is fat, so we invited Dr. Robert Serfass to come by and weigh the fat with us.

Ira: Dr. Serfass, how do you know if you're too fat?

Dr. Robert Serfass submerges Ira in a hot tub. Ira is hooked up to a scale that weighs him under water. Because fat floats (it is not as dense as water), the less Ira weighs under water the more fat he has in proportion to his total body weight.

NEWTON'S APPLE ■ 101

Dr. Serfass: There are several ways to determine that, Ira. The one most people are familiar with is the common height/weight chart. But the important question is whether you have a small frame, a medium frame or a large frame. You can determine that by taking skeletal measurements.

Ira: I like to think that I have a large frame.

Dr. Serfass: We all like to think we have a large frame because then we can weigh more. But if we take a measurement of your elbow width . . .

Ira: Why my elbow?

Dr. Serfass: It could be any number of bone measurements, and this is one of them. So we'll measure the width of your elbow, which turns out to be 2½ inches. Now, applying that measurement to our charts we see that, unfortunately, you have a small frame which means that your optimal weight is on the low side of that range.

Ira: Bad news. Well, what does that tell me about my fat? Am I a fat person?

Dr. Serfass: The limitation of height/weight charts is that they don't tell you what part of your body is fat, or how much of it is fat and how much of it is lean muscle tissue.

Another way of determining body fatness is to use skin-fold calipers to measure the fat stored underneath your skin. A lot of fat is stored just under the skin. So if we pull up two layers of skin in the front part of your wrist, we find two layers of skin with almost no fat. But we can look at other areas of the body, such as behind the arm, on the back, or on the stomach.

Ira: Let's leave my stomach out of this.

Dr. Serfass: OK. Let's take one here, just above your hip bone.

Ira: All of America has to see my love handles!

Dr. Serfass: This is a commonly measured site, and you have 15 millimeters of fat here.

By measuring several sites, as we did with you earlier, we can estimate that your body fat is about 17 percent.

Ira: Seventeen percent. Is that good or bad?

Dr. Serfass: That's not too bad. The average is 15 percent for your age group. You're well within reasonable limits of body fat.

Ira: Is that pretty accurate, then?

Dr. Serfass: It's a reasonably accurate method, but I've got an even better way of estimating your fat. You're really going to like this because you get a chance to dive into this hot tub.

Ira: Well, I really don't think I should dive!

Dr. Serfass: We've got a little boat for you to play with and lots of other things.

Fat is necessary to health and a normal part of our body makeup

---

THAT FAT

Most of the fat cells we have as adults were developed during childhood and adolescence. Most of the fat gained as an adult comes from increasing the size of those cells, not the number.

Ira: It's not every science show that you get to play in a hot tub. Carl Sagan, eat your heart out.

Dr. Serfass: An even more accurate way of determining your body fat is to measure your body density. Or, stated simply, the weight per unit volume of your body. Fat floats, but your bones and muscles sink. The less you weigh under water, the more fat you have because you are less dense under water. Now you have to let out all of the air in your lungs, lean over slowly, and submerge yourself completely under water.

The less you weigh under water, the more fat you have

Ira: All right, I'll give it a go.

Dr. Serfass: That's good. It's interesting that out of the water you weigh 157 pounds and in the water you weigh only 9½ pounds. Now we can compute your density and your percent of body fat, which with this underwater weighing method is 16 percent, not the 17 or 17½ percent found by the skin-fold method.

Ira: I certainly like this method better than the other one. Dr. Serfass, isn't there a certain kind of person who should worry about how much body fat he has in comparison with how much he weighs?

Dr. Serfass: Definitely. People who carry too much fat are more prone to suffer from conditions such as diabetes and hypertension, but you must remember that people can be overweight according to the charts and still not be "overfat." Hypertension, or high blood pressure, seems to be

Through the use of computers, however, the construction of an individualized diet is becoming a possibility. Computers can already analyze blood samples to find out how your body uses food. Soon computers

will look at a strand of your hair and tell you whether your body is deficient in certain trace minerals. They will also be able to examine your saliva and your gastric and enzyme juices to see how easily you can break down nutrients; moreover, the computer can study urine and fecal matter to find out what your body does and does not absorb. Even your chromosomes will be scanned for genetic deficiencies. The result will be a personal diet that can be modified as you age. In the meantime, *bon appétit*!

The optimum diet for every individual is not yet known

one of our nation's major obsessions, but not without reason. Each year a significant number of people die from it.

Ira: Do men and women have the same kind of fat problems?

Dr. Serfass: No, women have a little bit more fat than men do. The upper limit for reasonable fat for men is approximately 20 percent, whereas women can be up around 27 or 28 percent and still be within healthy limits.

Ira: People always wonder about when you start becoming fat. Is there a certain point in life when that is establshed?

Dr. Serfass: Much research suggests that most of your fat cells develop while you're growing. So it's important to take care of obesity when you're young. If you do it when you're young, you're not as likely to be overweight when you're older.

Ira: So someone like me, who was sort of skinny when I was a kid, might have a better chance of staying slim? I might be able to loose weight a little easier?

Dr. Serfass: Cut down on your food intake—lower the calories that you consume—and exercise. A combination of these is the best way to do it.

The writer Cyril Connolly once remarked that "Imprisoned in every fat man a thin one is wildly signaling to be let out." The ultimate goal of food nutritionists is to figure out the optimum diet for each individual so the thin person can emerge. There is certainly no shortage of dietary methods, but the optimum diet for every individual is not yet known. Scientists are just beginning to understand the biochemical nature of individuals and how to design personal diets geared to your age, health, genetic heritage, environment and work conditions.

A healthy body needs proper amounts of about forty chemicals, but no two people have the same requirements. Even two people of the same sex, height, weight and age can have substantially different basal metabolic (food-burning) rates. No single diet or FDA vitamin recommendation can completely satisfy everyone's needs.

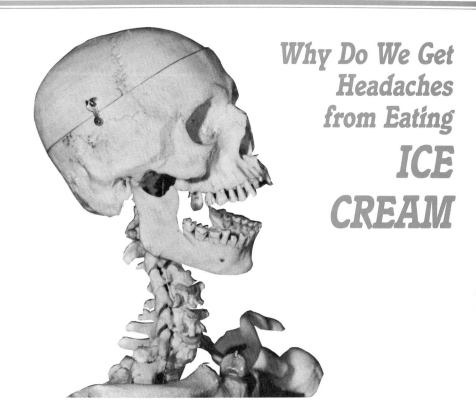

Why Do We Get Headaches from Eating ICE CREAM

Ira: When I eat ice cream fast—and I can eat it at lightning speed—my head hurts. Jan Serie, our expert on such weighty matters, is here to tell us why. Jan, I see you've brought along your undernourished friend.

Jan: Yes. I brought Dead Earnest with me, even though he doesn't eat much ice cream.

Ira: Bad for his diet, I suppose.

Jan: Right. It just drips off his mandible. But I thought I would bring him along to help explain what causes a headache when you put something cold in your mouth. There has been little research on this, so most answers are speculative. Some people think that if you put something cold in your mouth, like ice cream or an ice cube, it cools off your head. Then your body wants to send more blood to your head to keep it warm.

Ira: We're all hot heads, then?

Jan: Yes. To warm your head, the two car-

otid vessels that carry blood into the head expand so they can carry more blood. But they have pain receptors in their walls, and when the vessels enlarge, the pain receptors stretch, which brings on a headache.

Ira: So you're making up for the cold ice cream in your head by pumping more hot blood up there?

Jan: Right. One of the reasons many people accept this theory is because it explains migraine headaches.

Ira: Migraines? From eating ice cream?

Jan: No, you don't get migraines from eating ice cream, but they are caused by a dilation or an enlargement in the diameter of the blood vessels in the head. When the pain receptors on the surface of the blood vessels sense the dilation, migraines can occur.

Ira: Maybe I should figure out a way to make warm ice cream.

Jan: Good luck.

Parrots

Ira: A parrot is more than just a pretty talking bird, and here to prove that point is Nancy Gibson from the Minnesota Zoo. Welcome, Nancy, and who is your friend?
Nancy: This is Kohoutek, and he is a blue and gold macaw. Macaws are the largest members of the parrot family.
Ira: And where does he live normally?

Gorgeous colors help to camouflage the parrot

Nancy: Parrots live in the warm regions of the world. This one lives in Central and South America.
Ira: They have such beautiful colors!
Nancy: Yes, they have gorgeous colors. Even though it's hard to believe, these colors help to camouflage the bird.
Ira: This color? Blue? I think it would be easy for enemies to see these creatures.
Nancy: Well, against the blue sky and maybe against some yellow fruit they would have an ideal camouflage. Also, parrots see in colors, so they can learn to identify each other.
Ira: And this parrot has a big beak that I'm avoiding.
Nancy: It has a very big beak that can exert 300 pounds of pressure. Pretty impressive when you compare it with the mere 18 to 40 pounds of pressure in an adult male's grip. Let's demonstrate that beak and give him a walnut to crack.
Ira: He's going to break a walnut with his beak?
Nancy: See that long, round tongue that's coming out? Its surface is just like the tips of our fingers, textured to provide a good grip. He uses his tongue and beak to hold the walnut in place.

Another parrot that is with us is Alexander. He is an African grey from the western regions of Africa. Can you say hello?
Alexander: Hello.
Ira: He speaks very well.
Nancy: African greys are reputed to be the best talkers in the world of parrots.
Ira: Now if I bought a parrot in a pet shop, could I expect that bird to speak also?
Nancy: No. Let's clear up that misconception. For one thing, parrots don't talk, they mimic. They are truly the masters of mimicry. But only a few parrots will mimic sounds they hear around the house. Opinions differ as to why parrots mimic sounds they hear in the wild or in captivity. One theory is that, in the wild, they might mimic another bird in order to gain a larger territory. Or they might mimic a predator to scare it away. Another theory is that in the wild they are social and continually chatter back and forth, forming bonds in the flock. In captivity they can form a bond with the person who feeds them. So they mimic

Mimics without lips

sounds often heard to solidify the bond between them or to please the person.
Ira: I've always wondered how a parrot can form so many words and not have lips.
Nancy: They have a syrinx, which serves as their vocal organ. It's located at the bottom of the trachea, and there's where they form sound.
Ira: Whistling without lips. I'm envious.
Alexander: Bird noises.
Ira: I feel like a ventriloquist. How did birds come to have feathers?
Nancy: Birds evolved from reptiles.
Ira: Like snakes?
Nancy: Sure. You can see it in Alexander. He lacks a lot of feathers around his face and has a definite reptilian look. The feathers are modified scales.

Feathers that evolved from scales

Ira: These birds are beautiful and a lot of fun, but are they the kinds of birds you would take home as pets?
Nancy: No, no. I would like to discourage anyone from taking a parrot home as a pet. For several reasons. Only one out of 50 birds taken from the wild will survive long enough to make it as a pet. Not only are they expensive, but they live a long time and people tend to tire of them. Kohoutek is 20 years old and Alexander is 5. A bird like Kohoutek can live to be 80 or 90 years old. They say parrots have the intelligence of a three-year-old and the personality of a two-year-old.
Ira: Oh, the Terrible Twos!
Nancy: Can you imagine eighty or ninety years' worth of the Terrible Twos?

Tom Cajacob/Minnesota Zoological Garden

Behind the Scenes 10

NEWTON'S APPLE

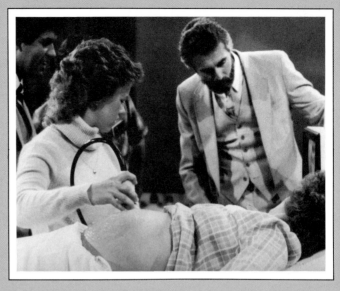

■ The pregnant woman in this segment is an obstetrician and gynecologist herself, and she had monitored many ultrasonic scans as part of her medical practice. Yet her excitement about seeing her own baby was infectious, and that baby acquired an extended family on the Newton's Apple set. ■ She and her husband wanted to know the sex of their baby, but they had a problem: *His* parents also wanted to know, but *hers* didn't. Luckily, the program wasn't broadcast until after the baby was born, so the secret was kept.

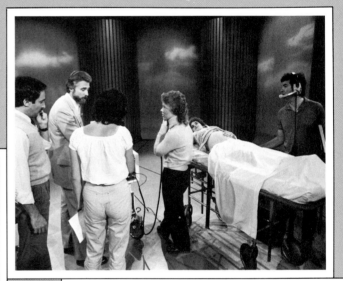

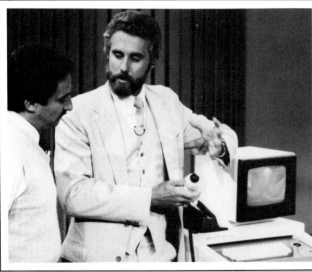

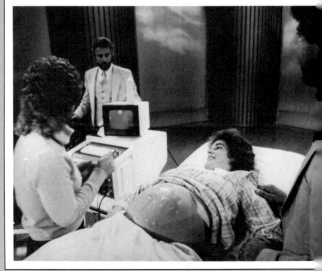

Space suits for the moon landing were cumbersome but life-sustaining; the shuttle astronauts will step out in suits that emphasize comfort.

DRESSING FOR SPA

by Robert G. Nichols

In the early days of space flight, astronauts wore modified U.S. Navy full-pressure suits as a backup to the spacecraft's pressurization system. These garments would enable astronauts to breathe if the capsule lost air pressure, but they could not keep them alive in the severe environment of space. So the first astronauts were confined to cramped quarters, awaiting a suit that would let them step outside.

The development of such a space suit was no simple task. In addition to supplying oxygen to astronauts leaving an orbiting spaceship, it must also protect them from the high-energy radiation of the sun. Moreover, the suit must prevent the astronaut from either frying or freezing in temperatures as high as 250 degrees Fahrenheit on the side exposed to the sun and as low as minus 65 degrees Fahrenheit on the side in shadows. And if these were not problems enough for one piece of clothing, a space suit must be able to withstand micrometeorites, tiny particles whipping through space. Such a collision has not yet occurred, but if it did and a micrometeorite punctured an astronaut's suit, oxygen would leak from the suit and the astronaut would need to get inside quickly.

A suit designed as part of the Gemini program solved these problems and allowed astronauts to emerge from the cocoon of their spacecraft to undertake extravehicular activity (EVA). This led to the development of a suit that permitted astronauts to walk on the surface of the moon. By the time of Skylab, astronauts were routinely leaving their orbital workshop to repair equipment, change film, and carry out a variety of EVAs.

The suit designed for space-shuttle missions will allow astronauts to do these jobs more readily and to perform other tasks that were never before possible. Technically referred to as the extravehicular mobility unit (EMU), the suit and life-support system provide astronauts with greater flexibility in their shoulders, arms and hands. When a Skylab astronaut tried to pull film from a telescope mounted outside the spacecraft, his movements were stiff and awkward, as though he were wearing twenty layers of binding clothing. Improved joints with bearings and friction-resistant material will allow shuttle astronauts to remove film from a satellite with greater facility. Shuttle astronauts will be handier in space—able to use a wrench more easily, for instance. Shuttle EMUs are not only superior in their performance characteristics, but in line with shrinking budgets at NASA, they are reusable and less expensive to maintain than previous suits. The days when a space suit would be used in one mission and then displayed in a museum are gone. Shuttle EMUs are designed to be used again and again over a fifteen-year period. Rather than producing each suit specially to fit each astronaut, NASA now manufactures components in sizes ranging from extra small to extra large. No longer the beneficiaries of custom tailoring, shuttle astronauts must shop off the shelf, choosing a pair of pants, for example, from a selection of six different sizes. When the mission is over, the suit is broken down into its various parts, cleaned and readied for reuse.

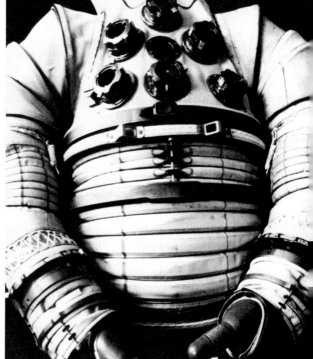

This prototype space suit was designed for the cancelled Apollo missions.

Anthony Wolff

Unlike previous space flights, where space suits were required attire, future shuttle flights call for astronauts to wear theirs only for extravehicular activity. Otherwise they will wear simple blue overalls. For an EVA, however, the astronaut first puts on a liquid cooling and ventilation garment that will dispose of the excess heat generated while working in a space suit. NASA quickly learned the importance of this article of clothing—several Gemini space walks had to be ended early because the astronauts became overheated and exhausted. Developed for use in the Apollo program, the liquid cooling and ventilation garment is a one-piece, zippered suit that looks like mesh long johns and contains about 300 feet of plastic tubing. By circulating cool water through this tubing, the garment can dispose of 2000 BTUs per hour, the amount of heat produced during strenuous exercise. The suit also contains hoses that collect the oxygen used to ventilate the body's surface and return it to the portable life-support system on the astronaut's back.

Once the liquid cooling and ventilation garment is on, the astronaut is ready to don the white outer suit that is familiar from earlier space flights. Improved design has significantly reduced the time astronauts need to spend dressing. It took an hour to put on the Apollo moon suits and life-support systems; it takes a shuttle astronaut 10 to 15 minutes to put on an EMU.

First, the astronaut dons the lower-torso assembly, which looks like a baggy pair of pants with attached boots. The lower-torso assembly is contructed of nine layers of material: the inner two, called the pressure-restraint garment, maintain the atmosphere inside the suit; the outer seven, called the thermal micrometeorite garment, protect the astronaut both from micrometeorites and the extreme temperatures of space. An astronaut puts on the lower-torso assemply much the same way one dons a pair of pants on earth—except in space you can put on your pants two legs at a time.

Second, the astronaut gets into the hard upper torso. This is a rigid, fiberglass shell that resembles a vest and has both sleeves and backpack life-support system attached to it. The hard upper torso is mounted to the air-lock wall of the vehicle. To put it on, the astronaut must squat down and then slide up, snaking the arms through the sleeves and putting the head out the neck ring. The astronaut then connects the upper and lower torsos by locking the waist ring.

Third, the astronaut puts on gloves that attach by wrist connectors to the sleeves. Though not custom-fitted, as they were for the Apollo suits, the EMU gloves are more flexible. Space-suit engineer Ronny Newman says that with the new EMU glove, an astronaut could pick up a dime—with practice. An insulating mitt like a kitchen hot pad can be placed over the glove when the astronaut must handle hot objects.

Finally, the astronaut puts on a helmet. Made of polycarbonate, a transparent impact-resistant plastic, this is one piece of equipment that has remained relatively unchanged since the days of Apollo. The visor assembly, with adjustable eyeshades to block out sunlight, is still placed over the helmet to protect it from damage. If the sunlight becomes excessively strong, the astronaut can pull down a special gold-plated visor that serves as a one-way mirror.

Anthony Wolff

Anthony Wolff

Anthony Wolff

Once inside the EMU, the astronaut checks out the crucial life-support system. The backpack is mounted to the suit so that all the oxygen and water hoses remain inside the hard upper torso. In this way no external hoses can become tangled or snagged. The system contains lithium hydroxide canisters that remove carbon dioxide so astronauts can rebreathe exhaled air. There is also a backup oxygen system that, if the primary system fails, can provide oxygen for 30 minutes, sufficient time for the astronaut to return to the shuttle orbiter.

The portable life-support system, which also houses the astronaut's radio communication equipment, has a chest-mounted display and control module. The astronaut can monitor and adjust life-support functions through the LED readout atop the module. The life-support system supplies enough power, water and oxygen to sustain an astronaut for up to seven hours. After checking pressure and oxygen flow, the astronaut is ready to disengage the EMU from the wall and depart the spaceship.

Because a flight plan might call for an EVA of as many as seven hours—a long time to be restricted to a space suit—NASA has provided the EMUs with some of the conveniences of home. If astronauts get thirsty, they can sip water from a tube in the lower part of the helmet that is connected to a half-liter drink bag mounted in the upper torso. If astronauts get hungry, they can eat a compressed fruit bar that is also positioned inside the helmet. If a male astronaut must urinate, there is a collection device under the liquid cooling and ventilation garment. (Plans call for women astronauts to wear a diaperlike Disposable Absorption Collection Trunk.)

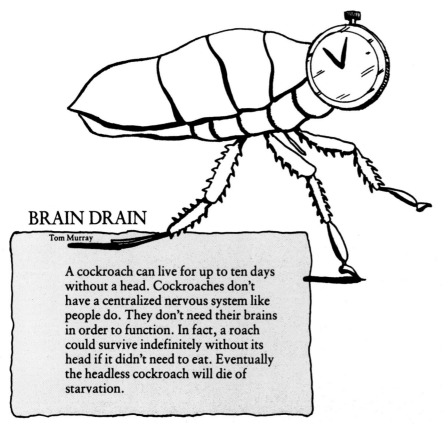

BRAIN DRAIN
Tom Murray

A cockroach can live for up to ten days without a head. Cockroaches don't have a centralized nervous system like people do. They don't need their brains in order to function. In fact, a roach could survive indefinitely without its head if it didn't need to eat. Eventually the headless cockroach will die of starvation.

Malfunctions in both life-support systems caused the cancellation of the EVA scheduled for the fifth shuttle mission last November. Both problems—a faulty magnetic sensor that disabled a blower fan and a small, plastic locking device left out of a regulator—were later diagnosed and corrected.

In the future, astronauts will don their EMUs and, attached to their orbiting craft by only a thin tether, perform important and exhilarating tasks in space.

(Reprinted with the permission of *Technology Illustrated* magazine. © 1983 by Technology Illustrated Publishing Corporation, 38 Commercial Wharf, Boston, MA 02110.)

Anthony Wolff

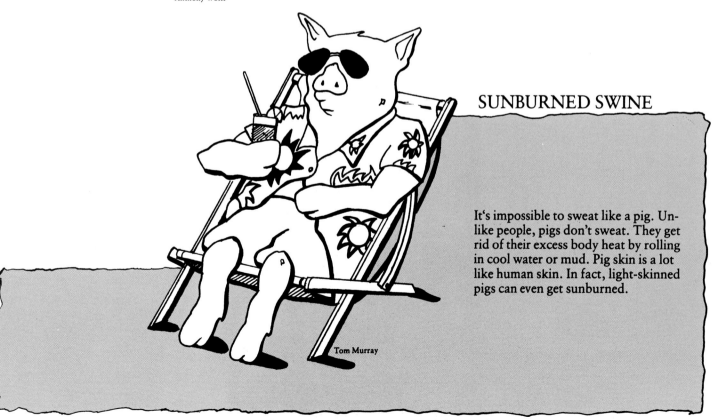

SUNBURNED SWINE

It's impossible to sweat like a pig. Unlike people, pigs don't sweat. They get rid of their excess body heat by rolling in cool water or mud. Pig skin is a lot like human skin. In fact, light-skinned pigs can even get sunburned.

Tom Murray

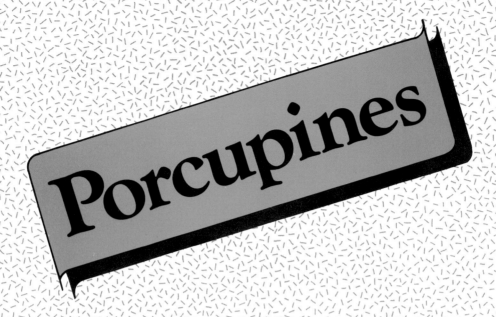

Porcupines

Ira: Welcome back, Nancy. I always thought porcupines were ugly, but the one you have with you is pretty.
Nancy: I think porcupines are regarded as ugly because people associate their painful quills with ugliness.
Ira: He's got an awful lot of them.
Nancy: He does. About 30,000 quills.
Ira: Are these all quills?
Nancy: No, some of these are just hair. The stiff white ones are quills.
Ira: And does he shoot them at people and predators?
Nancy: No, that's a common myth. Quills are just modified hairs. Porcupines can't shoot those quills any more than you can shoot your hair off your head.
Ira: How did that myth get started, then?
Nancy: There is some small truth to it. When they are in their defensive position, porcupines put their heads down and wag their tails back and forth. The quills are very loosely attached, so it looks like they are flying out, but they're really not. It's just the movement of the tail. To release them you have to touch them.

Tom Cajacob/Minnesota Zoological Garden

Ira: So if a predator sneaks up on it, this fellow will just back up or stand its ground?
Nancy: Yes. The base of the tail has the worst quills. You can see that they are a bit thicker there.
Ira: Where does he live in the wild?
Nancy: Most of them live in North America and Canada, where they spend a lot of time in trees.
Ira: What does a porcupine have to fear? With this kind of defense, what is going to attack it?
Nancy: They have several predators: lynxes, bobcats, foxes, coyotes. But the most successful predator of all is a little-known member of the weasel family called a fisher. He weighs about 10 pounds and he is very low to the ground. So he attacks the porcupine's head and flips him over to attack his belly, where there are no quills.

During birth the baby's quills are softened by the amniotic fluid

Ira: Speaking of quills, I don't know how to put this delicately, but if I were this porcupine's mother and I were giving birth to it, wouldn't I stab myself?
Nancy: Nature has taken care of that problem. An amniotic fluid in the birth canal softens the quills. The quills don't harden until they are exposed to the air for about three or four hours.

Tom Cajacob/Minnesota Zoological Garden

Ira: What happens when the predator is stuck with a quill?
Nancy: Most porcupine quills are barbed at the end. After you have been stabbed with a quill, the movements of your muscles guide the quills deeper into the flesh until a vital organ is pierced. It's a slow form of death.
Ira: That could take days, right?
Nancy: Yes, it could.
Ira: That's a terrible way to go. Are these animals rodents, Nancy?
Nancy: Yes. In fact they are the second largest rodent in North America. The beaver is the largest rodent.
Ira: What do they eat?
Nancy: Well, one of the distinctive characteristics of rodents is their constantly growing teeth, which means that they need to chew continually. These porcupines eat mostly tree bark.
Ira: I imagine that's unhealthy for the trees.
Nancy: It is, indeed. That's why it is good to have a stable population of predators for the porcupine. If you ever find these animals in the wild, be sure to give them their distance. They are attracted to campers because they like salt. The perspiration from human hands will cause them to go after canoe paddles or tools—things like that.
Ira: So if you wake up in the morning on a camping trip and your canoe paddles have been gnawed, you'll know what did it.
Nancy: That's right.
Ira: Well thank you, Nancy, for sticking it out with us here.

Canoe paddles or tools provide salty snacks

Super Sight with ULTRASOUND

Ultrasound waves are seen, not heard. Sound is, quite simply, the physiological sensation received by your ear. The source of any sound we hear vibrates somewhere between 20 and 20,500 hertz (a measurement indicating the frequency of a sound wave as it vibrates at one second per cycle); sounds beyond 20,500 hertz cannot be heard by human ears. It is almost inconceivable, but contemporary machines can generate millions of cycles per second—well beyond the range that can be heard by the most sensitive animals.

The wonder of ultrasonics is that it allows us to see and do things that only a few years ago were impossible. In the medical field alone, powerful diagnostic uses for ultrasound are becoming commonplace. In comparison with x rays, ultrasonics is a far safer method of medical testing.

To find out more about the medical uses for ultrasound, we invited Dr. Christopher Gostout to visit and show us some medical wizardry.

Ira: What better way to begin than with a baby in a mother's womb?

Dr. Gostout: Ira, we are going to use this ultrasound machine to make pictures of the baby. This probe emits and receives sound waves. Inside the probe is a piezoelectric crystal. When an electrical beam stimulates the crystal, it sends out a sound wave. This wave travels into the body and then bounces back. Finally it presents a video image. Different wavelengths of sound are interpreted by the machine in varying shades of black and white.

Ira: Like a TV picture.

Dr. Gostout: That's right. A gel is being smeared all over the mother's abdomen. We have to apply this acoustical gel, because if we don't the sound waves will scatter and we won't see anything.

Ira: What will you be looking for?

Dr. Gostout: Let's look at the head. A lot can be understood about the baby's development by examining the head.

Ira: Look at that! And now what are we looking at?

Dr. Gostout: The heart. That beating on this segment of the screen is the baby's heart. You can see the chambers of the heart. It's a fascinating thing to see.

The sound wave travels into the body and then bounces back

On the right is a sonogram of an infant in the womb. Notice the baby's head on the right and its arm circling across the top, with fingers curled and thumb extended. Sonograms help to recognize possible disorders and diseases. With this capability, doctors can sometimes diagnose and treat potentially fatal disorders before the baby is born. If necessary, it is possible to do some prenatal surgery.

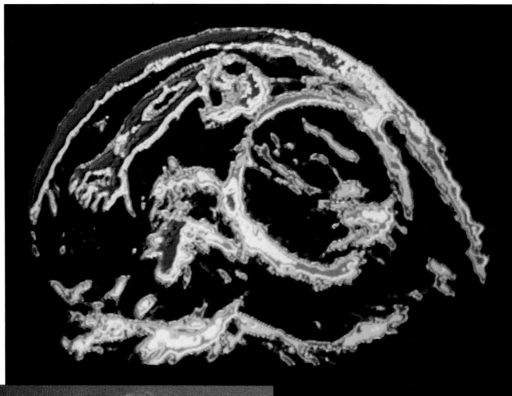

Howard Sochureck/Woodfin Camp

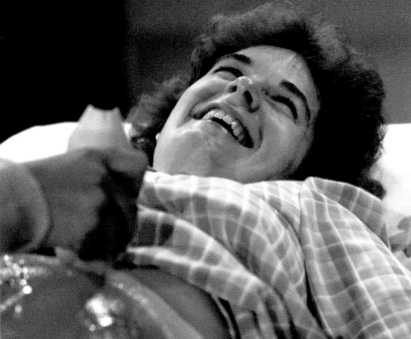

Dr. Gostout tells a happy mother that her baby is a boy. Ultrasound also helps determine the age of the fetus by allowing measurement of its skull. The gel on the mother's abdomen helps to conduct the sound waves.

Ira: Does the sound do any harm to Jan or the baby?
Dr. Gostout: No, it's in a frequency range that won't cause any harm. Now, with ultrasound we can sometimes identify the baby's sex. Jan, are you interested in finding out if you have a boy or girl?
Jan: I'd love to find out.
Ira: All right. Right here on national television. . . .
Dr. Gostout: It will take a little luck, but here goes. Look, we can see from this baby's genitals that it's a boy.

Revealing the baby's sex in the third trimester

Ira: Congratulations! I wish I had cigars to give out. How do you feel about that?
Jan: Wonderful!
Dr. Gostout: We can also look at the cardiology aspects of ultrasound by examining your chest, Ira. Let's bounce these sound waves off your heart. We'll put the probe just below your rib cage and . . . *voila*, we have Ira's heart. Cardiologists can see how efficiently your heart is functioning or

NEWTON'S APPLE ■ 115

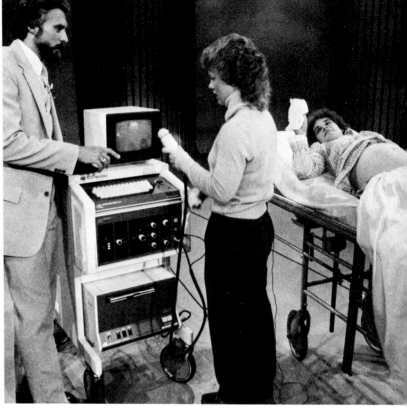

Ira bares his heart and Jan bares her abdomen for readings by the ultrasound machine. In the background an obstetrician waits to find out the sex of her own unborn child.

whether a valve or two may not be working properly.
Ira: And everybody's always saying I don't have a heart!

One of the common medical applications of ultrasound is the examining of the fetus within the mother's body. Ultrasound *scans*, or *sonograms*, are displayed on a screen to reveal the fetus' size, shape, sex, and possible disorders or diseases. Sonograms can detect whether the fetus is developing in the uterus or outside it—a condition that would threaten the lives of mother and fetus. Ultrasound can also help doctors to diagnose and treat fetal disorders such as respiratory or hormonal deficiencies.

Ultrasound devices are sensitive enough to capture the baby's profile in the thirteenth week of pregnancy, its internal organs in the twentieth week, and its genitals in the third trimester. Earlier in a pregnancy, ultrasound can observe how the mother's uterus changes even before an embryo is formed. Because of ultrasonics, the age of fetal care and prenatal surgery is upon us.

An ultrasonic method of measuring blood pressure has recently been developed.

SONAR SEARCH

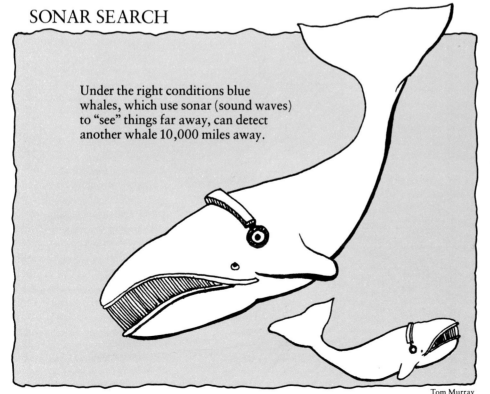

Under the right conditions blue whales, which use sonar (sound waves) to "see" things far away, can detect another whale 10,000 miles away.

Tom Murray

The sound beam is projected into the artery. Kortokoff sounds (a sharp thud, followed by swishing or blowing sounds) can be picked up in shocky patients or even newborn babies, when nurses and doctors can't hear anything in their stethoscopes.

A device more suited to the emergency room than to the home

The drawback of this blood pressure machine is that it requires a special ultrasound gel between the sensor and the skin, and the placement of the sensor must be very precise. It's a device more suited to an emergency room than to the home.

Ultrasound is being used in heart surgery. Until recently, to pinpoint the obstructed blood vessels doctors had to rely mainly on angiograms—x rays of vessels filled with a radiographic dye—taken prior to a coronary-bypass operation. But with angiograms it is very difficult to relate the pictures to what is actually seen during surgery. The ultrasound technique provides a way to look inside the arteries during the operation. Thus, surgeons can better judge where to place a bypass graft. By placing the ultrasound probe, which is about the size of a toothbrush, directly on a beating heart over the coronary arteries, a surgeon can distinguish between healthy and damaged arteries. In normal arteries the interior cavities of the heart appear black on the screen. But arteries blocked by fat and calcium appear to be filled with bright, globular structures that block (occlude) part of the interior cavities.

Breaking up kidney stones to be flushed away naturally

The apparent harmlessness of ultrasound is so appealing that it has gained popularity in probing all around the body: hearts, blood vessels, eyes, tumors, livers, kidneys and gall bladders. The kidneys, for example, can be rid of kidney stones with the help of ultrasonics. A device for conducting ultrasonic waves into kidney stones cracks them apart so they can be flushed from the body. The only drawback with ultrasonics seems to be that when air or other gases are present, such as in the lungs and intestines, the method is ineffective. Nor can it penetrate the skull.

Early applications of the ultrasonic generator, first developed by General Electric in 1949, included scrambling eggs, mixing paint, shattering glass, and homogenizing honey. Since then, the uses for ultrasound have burgeoned. It's good for cleaning small objects by vibrating them in a solvent, for echo sounding in deep water to map the topography, for massaging your muscles, and for agitating liquids to form emulsions. It can even detect hairline cracks in metals. As robots develop, ultrasonics will be a controlling sensory mechanism, as it is with many automatic cameras.

You can be sure you'll be hearing more about this!

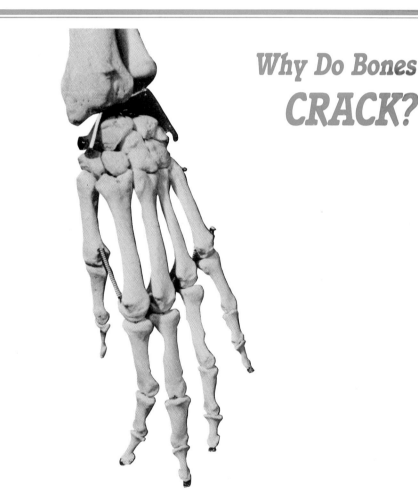

Why Do Bones CRACK?

Ira: It seems that my bones are always cracking, so we have invited Jan Serie and her bone-cracking friend, Dead Earnest, to tell us why.
Jan: I have Dead Earnest with me again because he's ideal for answering the question. He has lots of crackable bones.
Ira: Do they really crack? We don't actually crack bones, do we?
Jan: No, it's quite different from that. What we're really talking about is joints, which are between bones. That is, two bones opposed or adjacent to one another are separated by a cavity, the joint. If you pull on a joint, which contains a small amount of fluid, it responds like two suction cups being pulled apart. The popping sound results when you break the surface tension of the fluid in the joint.
Ira: No one is breaking any bones?
Jan: No bones at all.
Ira: How come you can't keep cracking your knuckles? If you do it once and then try to pop them again it won't work.
Jan: That's because it takes a while for the joint to reposition itself. There's also a different kind of joint cracking. Some joints can be cracked over and over. Every time you try, you can crack them. They have a knob or callous or bump in the way—an anatomical obstruction.
Ira: Thank you, Jan. And thank you, Dead Earnest, for cracking us up.

Publisher
General Communications Company of America
Jon B. Riffel, *Vice President and Director of Operations*
Robert H. Eggert, *Project Manager and Editor-in-Chief*

in association with
Twin Cities Public Television / KTCA
Richard O. Moore, *President and General Manager*
James E. Carufel, *Senior Vice President and Chief Financial Officer*
Timothy B. Conroy, *Vice President Development and Project Publications Executive*

Editorial and Production Services
BMR, *San Francisco*
Jack E. Jennings, *President*

Production Manager
Judy Johnstone

Writer
Howard Boyer
Jena Weiss, *Research Assistant*

Design and Art Direction
QR interface, *San Francisco*
Paul Quin, *President*
Jane Radosevic, *Operations*
Michael J. Patterson, *Mechanicals*

Photo Research
Travis Amos

Photography
Brian Paulson

Ira Flatow

Host of Newton's Apple, Ira has for twelve years been a science correspondent for National Public Radio, anchoring live remotes from such diverse locales as Three-Mile Island, Cape Canaveral and the South Pole. He has produced award-winning reports on science, health, medicine and technology, including the first live broadcasts from Antarctica since Admiral Byrd.

Nancy Gibson

A regular guest on Newton's Apple, Nancy is the Public Information Officer at the Minnesota Zoo, where she has been working with a variety of animals for four years. Nancy's love of animals and conservation have taken her to Central and South Americas, Africa, and into the tundra, to observe creatures in the wild.

Jan Serie

A Newton's Apple regular who often brings her friend Dead Earnest with her, Jan is an assistant professor of biology at Macalester College in St. Paul. She received her Ph.D. in anatomy in 1981 from the University of Minnesota. She is currently researching ways to transplant insulin-secreting tissue to reverse diabetes.

Host Writer
Ira Flatow

Producer
James Steinbach

Segment Producers
Sue Ballou
Lee Carey
Emily Goldberg

Associate Producer
Lynne Reeck

Photographer
Brian Paulson

Production Assistant
Debra Harper

Director
Bob Muens

Editor
Dan Luke

Production Supervisor
Jerry Huiting

Production Design
David Baumann

Lighting Design
Kendall Harris

Lighting Director
J. T. Price

Principal Audio
Chuck Preston

Principal Camera
Sylvia Chiu
Sheldon Erickson
Robert Hutchings

Principal Engineers
Tom Keleher
Jerry Lakso

Floor Director
Norbert Een

Vice President for Program Production
Jim Russell

Development and Marketing Services
Neil Dryburgh
Cal Thomas

Executive Producer
Gerald Richman

With thanks to:
Dr. Christopher Gostout, M.D., Nancy Gibson, Dr. Janet Serie, Dr. Lawrence Rudnick

And these people: Tom Adair, Bernard Beaudrey, Rich Brown, Peter Brownscombe, John Clouse, Scott Constans, Donna Dailey, Dianna Del Rosso, Robert Dorgan, Julie Francis, Gary Gaal, Alfonso Gallo, Johnny Hagen, Clayton D. Henderson, Susan Jones, Susan Woodruff Keane, Daniel Lutz, Michael Merrick, Bob Millea, Roy Misonznick, Valerie Mondor, Alan Moorman, Michael Phillips, Ric Reta, Steven Risenhoover, Kathleen Soulliere, Robert Sturm, Marilyn Timmsen, Jeffrey Weihe, Miles Wilkinson and Jeanne Young.

To order additional copies of NEWTON'S APPLE or receive information about educational discounts for quantity purchases, write:

NEWTON'S APPLE
Business Management Research
1668 Lombard Street
San Francisco, California 94123